INSIDE YOUR
CALCULATOR

THE WILEY BICENTENNIAL–KNOWLEDGE FOR GENERATIONS

*E*ach generation has its unique needs and aspirations. When Charles Wiley first opened his small printing shop in lower Manhattan in 1807, it was a generation of boundless potential searching for an identity. And we were there, helping to define a new American literary tradition. Over half a century later, in the midst of the Second Industrial Revolution, it was a generation focused on building the future. Once again, we were there, supplying the critical scientific, technical, and engineering knowledge that helped frame the world. Throughout the 20th Century, and into the new millennium, nations began to reach out beyond their own borders and a new international community was born. Wiley was there, expanding its operations around the world to enable a global exchange of ideas, opinions, and know-how.

For 200 years, Wiley has been an integral part of each generation's journey, enabling the flow of information and understanding necessary to meet their needs and fulfill their aspirations. Today, bold new technologies are changing the way we live and learn. Wiley will be there, providing you the must-have knowledge you need to imagine new worlds, new possibilities, and new opportunities.

Generations come and go, but you can always count on Wiley to provide you the knowledge you need, when and where you need it!

WILLIAM J. PESCE
PRESIDENT AND CHIEF EXECUTIVE OFFICER

PETER BOOTH WILEY
CHAIRMAN OF THE BOARD

INSIDE YOUR CALCULATOR

From Simple Programs to Significant Insights

GERALD R. RISING
State University of New York Distinguished Teaching Professor
University at Buffalo
Buffalo, New York

WILEY-
INTERSCIENCE

A JOHN WILEY & SONS, INC., PUBLICATION

Published by John Wiley & Sons, Inc., Hoboken, New Jersey.
Published simultaneously in Canada.

For general information on our other products and services or for technical support, please contact our Customer Care Department within the United States at (800) 762-2974, outside the United States at (317) 572-3993 or fax (317) 572-4002.

Wiley also publishes its books in a variety of electronic formats. Some content that appears in print may not be available in electronic formats. For more information about Wiley products, visit our web site at www.wiley.com.

Wiley Bicentennial Logo: Richard J. Pacifico

Library of Congress Cataloging-in-Publication Data is available.

Rising, Gerald R.
  Inside your calculator : from simple programs to significant insights Gerald R. Rising.
      p. cm.
  Includes bibliographical references and index.
  ISBN: 978-0-470-11401-8
  [1. Calculators. 2. Mathematical instruments. 3. Office equipment and supplies.] I. Title
  QA75.R57 2007
  510.28'4–dc22

                                                            2006052567

Printed in the United States of America

10 9 8 7 6 5 4 3 2 1

To

Don Stover

Teacher, scholar, friend

CONTENTS

PREFACE

Computation is all about shortcuts.

Consider, for example, how you would compute $4 * 5$.[1]

A reasonable response would be "I don't 'compute' anything. I simply recall a fact I have memorized: $4 * 5 = 20$."

Indeed, that is how almost all of us arrive at that answer now, but consider how we got to this point historically. That $4 * 5$ is a shortcut notation for adding four 5s, that is, $4 * 5 = 5 + 5 + 5 + 5$ (or equivalently five 4s), and multiplication of whole numbers in general is simply shortcut addition.

Thus we could solve the original exercise $4 * 5$ by adding four 5s, again applying memorized "facts": $5 + 5 = 10$, $10 + 5 = 15$, and finally $15 + 5 = 20$, to arrive at our final answer.

But addition is also a shortcut, in this case a shortcut for counting. Adding $5 + 5$ is, for remote uncivilized tribes as well as for small children, a kind of "counting on." We can calculate it by starting with the first 5 and counting five more places, thus:

$$5 + 5$$

$$1, 2, 3, 4, 5, \quad 6, 7, 8, 9, 10$$

---

[1]Throughout this book the now standard computer science symbol $*$ represents multiplication.

So our original calculation has been reduced to counting:

$4 * 5$

$5 + 5 + 5 + 5$

first 5 + second 5 + third 5 + fourth 5

$1, 2, 3, 4, 5,$ $6, 7, 8, 9, 10,$ $11, 12, 13, 14, 15,$ $16, 17, 18, 19, 20$

Thus again: computation is all about shortcuts.

But clearly from these examples, learning math is also about procedures and memorization. We needed a procedure to change counting to addition and another to change addition to multiplication, and, of course, we'll need many more procedures—to change multiplication to exponentiation and to calculate with fractions, for example—to carry out more and more sophisticated computations. Along the way we need to memorize not only those procedures (called *algorithms*) but also those addition and multiplication facts, like $5 + 6 = 11$ and $4 * 5 = 20$, in order to carry out computations efficiently.

It is these last two components of computation that lead us quite naturally to computing devices. What calculators contribute is a great deal of memory and a great deal of speed. Even the cheapest handheld calculator operates so rapidly that it could compute $4 * 5$ by counting about as fast as we could recall the fact that we know is the answer. They don't have to do that, however. Engineers have "taught" our calculators just as parents and teachers have taught our children to take shortcuts in computation. And it is exactly those kinds of shortcuts that are described in this book.

What is so remarkable about those shortcuts is how elementary they are. To understand them, all that is required of the reader is a general knowledge of school arithmetic and algebra, which, for those who have forgotten some of that, will be reviewed along the way.

You will see in this book how the various calculator keys can be backed up by procedures—how, that is, your \$15 scientific calculator can calculate

$$(1.05)^{400} = 299033351.2$$

$$\sqrt{37} = 6.08276253$$

$$7^{.37} = 2.054406$$

$$\log 387 = 2.587710965$$

$$\cos 108° = -0.3090169944$$

$$\sin^{-1}(.9440057250) = 1.23456789$$

and does so as rapidly as you can key in the numbers and the operation key.

Two reservations must be recognized. Although the programs introduced in this book will in most cases calculate answers about as rapidly
as will the hardwired procedures in the calculator, *they are seldom the
same procedures*. In order to gain efficiency, those manufacturers' algorithms draw upon more complicated techniques. They are also carefully
guarded trade secrets. You will, however, meet a simulation of one of
those techniques in Chapter 9.

I also note that few of the concepts of this text are original with me.
Most are drawn, with permission, from a remarkable book in preparation,
Donald Stover's *PreCalculus Problems and Projects*. My role in this book
is still, I suggest, an honorable one: interpreting those important algorithms
and programs for a wider audience. I hope that your reading will prove
me successful in that endeavor.

ON READING THIS BOOK

This book is aimed at a range of audiences. I suggest here how different
readers might approach the contents.

General Readers

Many of you will not have been using mathematics regularly, and your
memory of details may be less than you would like. I urge you to read this
book straight through, because I summarize any necessary mathematics
background with care. I think that you will take from this approach a
better understanding of topics you only thought you mastered as a school
student.

It will not be necessary for you to follow the programs in detail or
to program a calculator yourself in order to see how the problems are
addressed. But don't simply skip the programs; they convey information
as well.

I also suggest that you read Appendix C between reading Chapters 1
and 2. It will provide further background for contemporary calculation.

Readers Interested in Programming

Whether or not you are an experienced programmer, you should find the
programs of this book instructive. If you have one of the Texas Instruments
TI-84 or TI-83 series calculators, you can key in the programs and see
how they work. I urge you, however, not simply to copy the programs.
Be concerned with their structure to see how they work, for many of the
ideas introduced are also broadly applicable.

Appendix A should answer any questions you have about TI-84 programming details. For users of other calculators, a supplementary Appendix A appropriate for other Texas Instruments calculators as well as those of the Casio FX series is available for free download from the website:

`www.buffalo.edu/~insrisg/InsideYourCalculator/`

Other reference and supplementary materials are also available from this website.

Again I suggest that you read the book straight through because Chapter 1 will remind you of some of the benefits of modern calculation, while Chapter 2 develops some important ideas that support subsequent programming.

You should also be interested in the special techniques presented in Appendixes E, I, J, K, L, and M.

Readers Interested in Specific Topics

This book should serve as a useful reference in support of individual topics. If you are interested in logarithms, for example, turn to that chapter. There you will find almost all of the presentation related to that topic, independent of the rest of the book. In particular, if you are a teacher, you should find ways to extend and enliven your presentation of such topics by using this approach.

Chapter 9 is special in this regard, for it provides some insight into the CORDIC program that supports so much of what goes on in calculating machines, including sophisticated computers as well as the simplest handheld scientific calculators.

All Readers

As with most authors, I have found the years preparing this book a happy experience. I hope that I convey my enjoyment working with these ideas and that you, too, will gain from considering them seriously.

GERALD R. RISING

Buffalo, New York
March 2007

PART I

THE SETTING

1

INTRODUCTION

"Reeling and Writing, of course, to begin with," the Mock Turtle replied,
"and the different branches of Arithmetic—
Ambition, Distraction, Uglification and Derision."

—Lewis Carroll

The word *algorithm* is central to mathematics and computer science. It means a step-by-step problem solving procedure. All of us have learned many algorithms in our schooling. The algorithm for "long division" is a good example. I still recall the oversimplified version we students pronounced as a refrain for our fourth-grade teacher: "divide, multiply, subtract, bring down, repeat." It never occurred to that teacher—or to us students, of course—that it might be appropriate for us to understand what was going on as we carried out what were to us a series of mindless rote activities.

We should always have been concerned with why algorithms like that work. Fortunately, the computer revolution has brought that concern to the forefront.

The seeds of my own investigation of algorithms were planted at a math meeting in Kansas City in 1972. A Hewlett-Packard Company representative who had attended a talk I gave on computation caught my attention and introduced himself. At the time that company was producing some of the early desktop electronic calculators, so I was pleased to talk with him.

Figure 1.1   The HP-35 scientific calculator of the 1970s.

"I've got something to show you," he told me slyly, the twinkle in his eye reminding me of those grifters whose inside pockets are filled with stolen wristwatches. He did not, however, have contraband for sale. Instead he took from his jacket pocket a small leather-encased parcel. He opened it to disclose the first handheld scientific calculator, an HP-35 (Figure 1.1).

As I write this over 30 years later, I find it difficult to communicate the astonishment I felt on that morning. Until that time the only electronic calculators available were "four-bangers"—so called because their processing was limited to the four fundamental operations of paper-and-pencil arithmetic: addition, subtraction, multiplication, and division. Even those had not been around for long. The first handheld calculators had been available only since 1970, the first electronic desktop calculators since 1963. Here was a calculator that not only performed those four operations but also—for the first time—calculated trigonometric and logarithmic functions, reciprocals, and roots.

Still more impressive to me, when I punched the appropriate keys to enter $2.356^{3.71}$, the calculator almost instantaneously displayed 24.03091523.

To gain some sense of both my astonishment and the remarkable power this tiny instrument provided, consider how I would have had to address that problem at that time. (Logarithms will be reviewed later, and you do not have to follow the details of this worked example to understand my point.)

I would have written the exercise as an equation $x = 2.356^{3.71}$, then taken the logarithm of each side, in the process applying one of the rules of logs to the right side:[1] $\log x = 3.71 * \log 2.356$.

Next I would have looked up log 2.356 in a (base 10) log table, interpolating[2] to give .3722, annexed the appropriate characteristic, 0, and substituted it in that equation: $\log x = 3.71 * 0.3722$.

Now I would have multiplied those right-side factors with a simple calculator[3] to produce log $x = 1.3809$.

Finally, I would have returned to log tables to find, by interpolating again, the antilog of 1.3809 to arrive finally at $x = 24.04$.

Notice several things about that processing. First, of course, it was lengthy and time-consuming—and I have not even included the interpolation procedures. Moreover, it gave nowhere near the number of decimal places of the calculator answer; and finally, the answer it did produce was not even accurate to that fourth digit.[4]

Knowing all this, I was stunned. A single exercise that would have taken me at least 5 minutes was now calculated as fast as I could key in the numbers and operation. Electronic engineers had been able to pack into this tiny device tremendous computing power, and I could not imagine how they performed this feat of calculating wizardry.

It turns out, I now learn, that even the manufacture of this calculator was a kind of fluke. When one of the early electronic desktop calculators was developed that would compute with this power, William Hewlett, the head of Hewlett-Packard Company, was impressed with the small space taken up inside the case by its electronic components. He asked his engineers if they could squeeze this power into a shirtpocket-sized calculator. The engineers responded with the HP-35. At first they planned to make only a few: for their boss, other company administrators, their engineering colleagues, and, of course, themselves. Fortunately for

---

[1] A reminder: throughout this book, * represents multiplication.

[2] Sadly, one of the "benefits" for calculators that has attracted some teachers to them is that their extra digits "eliminate the need for interpolation." See Appendix B for more on this important mathematical technique.

[3] Without an electronic calculator that processed arithmetic, I would have had to choose between multiplying those numbers by paper and pencil or taking logs again, complicating the calculation still further.

[4] Some older readers may recall that a slide rule would have simplified matters for those who knew how to use one, but even less accuracy would have been possible: to three digits on all but a few expensive models that provided four.

them they changed their minds because, by the time this calculator was finally outmoded a few years later, tens of thousands had been sold—for $395 each!

Of course, within a few years Japanese manufacturers would flood the market with inexpensive four-bangers and scientific calculators. At the height of those times when—as what was called Japan, Inc.—our defeated World War II opponent seemed about to reverse the results of that war economically, calculator prices reached rock bottom. Even programmable calculators could then be purchased for under $10.

But at that meeting my imagination was captured by that tiny instrument. I wanted to know what was happening inside that calculator. How did it magically produce those results? How did those engineers accomplish this further advance in computation?

I knew that electronic engineers had several things going for them that their predecessors did not have. Their electronics gave them:

1. Great computational speed
2. Large storage (memory) capacity
3. Programming opportunities

Somehow they had harnessed those electronic gifts to produce what were to me and others such brilliant results.

THE BLACK BOX

The problem I faced, it seemed to me, was related to a pedagogical device that is useful for encouraging students at almost any level to think seriously about mathematics. The teaching device is often called a *black box* or *function machine*, and the challenge is *What's My Rule?*

The students are presented with an imaginary black box into which you can feed a number. Each time you enter a number you receive in response a corresponding answer number. By testing with as many input numbers as you wish, you are asked to determine what is happening inside the black box, what mathematical operations are doing to that input number to manufacture that output number.

The students don't even have to state the rule in words. By showing their teacher that they can correctly predict what output number results from the input numbers with which they are challenged, they demonstrate that they know "what is going on in the box." (This way they also don't disclose the secret to others who can continue to seek the rule.)

Figure 1.2   A function "machine."

Consider a sample black box problem. You are to determine what the following box does to numbers dropped in the top funnel. Shown in Figure 1.2 is a 7 entered to produce 33.[5]

By trying other input values, students would seek to determine the rule that would produce these results. Mathematicians recognize such boxes as the equivalent of functions.

I, too, sought to determine what is going on in a black box, except that my box was equivalent to a scientific calculator key. Also my search was different from the search in *What's My Rule?* I know the rule; it is stamped on or near that key. What I sought and what this book is about is how that rule might be accomplished—for example, what could be going on inside that black box labeled COS (see Figure 1.3).

In this task trial and error will not suffice, however. You can enter value after value to obtain outcome after outcome without making much progress in determining what is going on inside the box. So a quite different approach is required. In order to answer this question you have to explore the mathematics of the function cosine[6] as well as the programming necessary to support that math. And that is what you will do in the remainder of this book.

---

[5]Without further information, there is a wide range of possibilities for this box rule. If we consider the input value as x, the rule could be $5x - 2$ or $x^2 - 16$ or even just 33 for every input x. Mathematicians know that the rule for a finite number of specific input values need not be unique, but that does not affect the game as it is played with less sophisticated contestants.

[6]I have arbitrarily chosen cosine, abbreviated COS on the calculator, to represent one of the circular or trigonometric functions. Once we have the means for calculating its values, the sine, SIN, and tangent, TAN, keys are quickly determined by use of trigonometric identities.

149

cos

−.85716730074

Figure 1.3   The COS key as a black box.

SUPERHUMAN ENGINEERS?

I have always held engineers in very high regard. They are the "Can do!" people of this world. Given a practical problem, they set out to solve it. Build a dam, erect a skyscraper, construct a road, send a rocket to the moon: they get at it. I honor them for their creativity and their work ethic.[7] But at first the awe I felt for engineers got in the way of my figuring out how computers calculate. I was certain that they were applying some very advanced and highly abstruse math in extraordinarily complex programs to solve these problems.

That it turned out otherwise came as a revelation to me as I hope it will to you. In the following chapters you will meet simple programs that carry out the functions of those calculator keys. In the process you should gain further insights into the mathematical and programming concepts that support them, insights that should serve you well in other contexts. And the number of program steps needed to carry out these tasks is several orders of magnitude fewer than those that support contemporary computer games.

Here, for example, is a seven-line program that will calculate the cosine for the input of any number of degrees to nine- or ten-digit accuracy:[8]

---

[7]Basing my judgment on my over 40 years of university lecturing, I prefer teaching classes of engineers to all other students.
[8]Those who wish to enter this program should find Appendix A on programming specifics useful.

```
PROGRAM:COSDEG
: Prompt X
: X*π/180→X
: X*X/4294967296→S
: For (I,1,16)
:     S(4-S)→S
: End (For)
: Disp 1-S/2
```

If you enter those seven program lines in a programmable calculator and run the program, it will carry out this seemingly formidable task.

Suppose now, for example, that you wish to calculate the cosine of 149°. When you run the program, the calculator will display X=?, to which you would respond by keying 149 and pressing ENTER.[9] Your display would then look like this:

```
X=?149
       -.8571673008
               Done
```

You will see how and why that remarkable program works in Chapter 8. For now I want only to show you the tasks those steps are performing:

Prompt X   This is the program line that displays that X? when the program is run, inviting you to type in a number of degrees.

X*π/180→X   In this line the number you entered in response to the prompt in line 1 is multiplied by π. The value of π (3.1415926535898) is stored[10] in calculator memory. The product you attain is then divided by 180. The arrow tells you to store this result in X, replacing any value that was there previously.

X*X/4294967296→S   More arithmetic. Our new stored X is multiplied by itself and then divided by that strange number, which looks

---

[9]Although it is not necessary in order to follow the arguments in this book, I strongly encourage you to run this and other programs. To run this program it is not necessary to set your calculator in Degree mode, but to check it you would need to do so.

[10]Most calculators will show stored values like π to the limited number of digits of the display, but will carry more digits in memory. To check any calculator to determine how many digits of π are stored, enter π, subtract 3.1415926, and multiply by 10000000. If your calculator value for π had been the 3.141592654 displayed, the result of this calculation would have been .54. When you see something different from this, in the case of the TI-84 .535898, those digits replace the 54 to give us 3.1415926535898. Thus this calculator carries π accurate to 14 digits. Other calculators and computers will differ, and it is an interesting task to check them out, not just for π but for calculated values such as $\sqrt{2}$ as well.

like the national debt and happens to be 2^{32}. The quotient, clearly a very small number, is stored in S.

```
For (I,1,16)
S(4-S)→S
End
```
These three lines form what is called a *counting loop*. In the For line an internal counter I runs the lines between it and the End line 16 times. (It "counts" from the first value, 1, to 16.) Of course, here there is only one line to be calculated over and over. It takes the current value of S, multiplies it by 4 minus that same value, and stores the result back in S. (You can think of the For statement as "For I taking values from 1 to 16, do the following:.")

The power of this control stucture is displayed clearly here. These three lines replace 16 program lines, all alike:

```
S(4-S)→S
S(4-S)→S
...
S(4-S)→S
```

Disp 1−S/2   One last minor bit of arithmetic. The final S result produced by that For loop is divided by 2 and the result subtracted from 1. The answer is then displayed. It is the cosine of your original input X, in our example 149°.

To understand what went on in those steps, readers not already familiar with simple programming will have had to learn from this analysis about input (Prompt), output (Disp), and calculator storage (→) and how a particular loop (For) works. Aside from those programming features, however, all that is involved here is simple arithmetic, in this case subtraction, multiplication, and division. Of course, that means simple for the calculator! None of us would want to divide by 4294967296 or carry out even one of those 16 multiplications of 10-digit factors without access to such a device.

Before closing this discussion, I must enter an important reservation. Some of you will have checked the result of our program calculation of cosine 149°. Our result, −.8571673008, does not quite agree with the result you obtain when you simply use the scientific calculator keyboard to key $\boxed{\text{COS}}$ 149 and press $\boxed{\text{ENTER}}$. If you do that, the calculator will display − .8571673007,[11] which differs by one in that tenth place from

---

[11]If you do that and get the wildly different answer, −.2237409501, your calculator is in Radian mode and you must change it to Degree mode.

our program calculation. An error of that magnitude corresponds to a measurement that is off by less than an inch in 100,000 miles, but the two values do differ, suggesting that they are arrived at by different internal processing avenues.

In fact, cosine is calculated by most computers and calculators by a quite different (and still faster) program. You will meet a simulation of that program in Chapter 9—not, however, to make this tiny correction but only because it involves additional interesting mathematics and programming.

JUST HOW POWERFUL IS THAT PROGRAM?

I invite you to examine the power of that seven-line program by comparing it with what you would have needed to do to accomplish the same result before the advent of electronic calculation.[12]

Consider Table 1.1, which gives only the values for angles in whole-number degrees. If you wished to give values for tenths of a degree ($1.0°$, $1.1°$, $1.2°$, etc.), the table would have to be 10 times as long. Similarly, for values to hundredths of a degree ($1.01°$, $1.02°$, $1.03°$, etc.), it would have to be 100 times as long.

But our seven-line program provides values for angles measured to hundred millionths of a degree. For example, our scientific calculator tells us that $\cos 45.12345678° = .0705581518$. To provide all this information (even allowing for linear interpolation), a table 10,000,000 times as long as Table 1.1 would be necessary. Whereas Table 1.1, in degrees, takes up one page, a table that would provide all this information would call for 10 million pages. That is, of course, a great many pages. It would take twenty thousand 500-page volumes to include all of them, an entire library devoted to the values corresponding to this one of the many calculations that our little calculators so simply programmed can perform.

Two reasonable questions arise at this point. First, angles are rarely given in decimal values. Instead, like hours, degrees are broken down into 60 minutes and each minute into 60 seconds. For example, you might have an angle of $27°39'12''$, that notation representing 27 degrees, 39 minutes, and 12 seconds. While some calculators allow input in this form, many do not. It is reasonably simple to convert from one form to the other,

---

[12]This discussion is oversimplified. More information could be included on each page, and various shortcuts are employed by books of tables to reduce the number of pages. However, the message remains: a remarkable amount of information may be retrieved through calculator processing.

TABLE 1.1. Cosine Values from 0° to 90°

Degrees	Cosine	Degrees	Cosine
0	1.000000000	45	.707106781
1	.999847695	46	.694658370
2	.999390827	47	.681998360
3	.998629535	48	.669130606
4	.997564050	49	.656059029
5	.996194698	50	.642787610
6	.994521895	51	.629320391
7	.992546152	52	.615661475
8	.990268069	53	.601815023
9	.987688341	54	.587785252
10	.984807753	55	.573576436
11	.981627183	56	.559192903
12	.978147601	57	.544639035
13	.974370065	58	.529919264
14	.970295726	59	.515038075
15	.965925826	60	.500000000
16	.961261696	61	.484809620
17	.956304756	62	.469471563
18	.951056516	63	.453990500
19	.945518576	64	.438371147
20	.939692621	65	.422618262
21	.933580426	66	.406736643
22	.927183855	67	.390731128
23	.920504853	68	.374606593
24	.913545458	69	.358367950
25	.906307787	70	.342020143
26	.898794046	71	.325568154
27	.891006524	72	.309016994
28	.882947593	73	.292371705
29	.874619707	74	.275637356
30	.866025404	75	.258819045
31	.857167301	76	.241921896
32	.848048096	77	.224951054
33	.838670568	78	.207911691
34	.829037573	79	.190808995
35	.819152044	80	.173648178
36	.809016994	81	.156434465
37	.798635510	82	.139173101
38	.788010754	83	.121869343
39	.777145961	84	.104528463
40	.766044443	85	.087155743
41	.754709580	86	.069756474
42	.743144825	87	.052335956
43	.731353702	88	.034899497
44	.719339800	89	.017452406
45	.707106781	90	.000000000

however. There are 60 minutes in a degree and 60 times 60 seconds (thus 3600 seconds) in a degree, so you calculate

$$27^\circ 39' 12'' = \left(27 + \frac{39}{60} + \frac{12}{3600}\right)^\circ = 27.65333333^\circ$$

and find the cosine of this result. This gives $\cos 27^\circ 39' 12'' = .8857719399$.

The second question deserves a serious response. Is all this accuracy of any real value? Cannot you get along just as well with the four- or five-digit accuracy of those older days?

Mathematics, science, and engineering teachers at all levels should be especially sensitive to such questions for they face the obverse of this problem: their students are quite content with meaningless accuracy. When asked to find the circumference of a circle with a 12.0 inch diameter,[13] for example, those youngsters who know the formula $C = \pi d$ will enter 12 into a calculator with 10-digit display and multiply it by the calculator value of π to obtain 37.69911184. They are quite satisfied then to give 37.69911184 inches as their answer when most of those decimal digits unnecessarily confound the problem. A more appropriate answer would be 37.7 inches.[14]

But we are still faced with this fair inquiry: Is all this extra accuracy ever of any value?

An example should respond to this question. Global Positioning System satellite (GPS) devices (e.g., see Figure 1.4) are widely used today. GPS instruments, formerly equipment restricted to the armed forces, have become widely available. They are used by surveyors; travelers by airplane, car, and boat; hunters and explorers; and parcel delivery personnel. Although these tools, the size of handheld calculators, provide many other features—most notably maps—their basic function is to determine your location on the earth; that is, your latitude and longitude. This is accomplished by calculating your position in relation to a number of satellites.

When I take my GPS out in my backyard and turn it on, it reports my latitude as $44^\circ 00.179'$ north, my longitude as $78^\circ 44.932'$ west. It also reports how accurate these values are, this accuracy depending on the

---

[13]The measure 12.0 inches differs from the measure 12 inches. The measure 12.0 inches indicates that the measure is to the nearest tenth of an inch; the measure 12, to the nearest inch. In formal terms, if a measurement m is 12.0, then $11.95 \le m < 12.05$. If, on the other hand, a measurement n is 12, then $11.5 \le n < 12.5$.

[14]There are rules governing significant digits obtained from calculations with numbers arrived at through measurement. Note that in this case we have simply retained the same number of digits—three—as the given diameter 12.0. It should also be noted that there is an exact, but abstract and not useful for measurement, answer to this question: 12π.

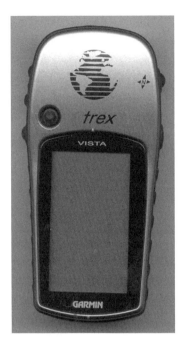

Figure 1.4   A GPS device.

number of satellites it can "see" from this location.[15] When I took that reading, for example, my GPS reported "Accurate to 19 feet."

Even readers unfamiliar with GPS devices have almost surely seen their contribution under less happy circumstances. Those laser-guided "smart" bombs employed by armed forces that miss their targets by at most a few feet have similar tools built into them. World War II veterans like me are especially impressed by that accuracy. During that war just 60 years ago, the proportion of bombs dropped by our B-17 Flying Fortresses over Europe that fell within 1000 feet of a designated target was only 20%. Those were daylight raids; the British nighttime bombing was still less accurate.[16]

If those examples don't make the case for that many digits, you need only extend these kinds of tasks to the accurate location of spacecraft as they travel through the solar system and beyond. An idea of the kind

[15]Under a heavy canopy of trees, GPS devices are of little use. Thus open areas and winter provide better accuracy.
[16]This information is derived from page 5 of Franklin D'Olier et al., *The United States Strategic Bombing Survey Summary Report* (European War), available from www.anesi.com/ussbs02.htm.

of accuracy called for in astronomical measurements is suggested by the definition of the meter to be the length of a path traveled by light through a vacuum in .00000003335640952 second.

Where does the remarkable precision of these contemporary devices come from? It should be evident that one source of this precision is our ability to calculate to extreme accuracy.

WHAT LIES AHEAD

In the remainder of this book I will share with you the results of my exploration of electronic calculation. Included will be the basis for that cosine program and the other features that make the scientific calculator so different from the four-bangers that came before them.

But first we must set the stage for modern electronic calculation. In Chapter 2 you will meet some additional mathematical background to support this history. Then you will examine some remarkable algorithms that could be used to back up the calculator keys $\boxed{\sqrt{}}$, $\boxed{\cos}$, $\boxed{\log}$ and $\boxed{x^y}$. (General readers may wish to review the history preceding electronic calculation in Appendix C before continuing with Chapter 2.)

PART II

ALGORITHMS AND PROGRAMS

2

NUMBERS, ALGORITHMS, AND PROGRAMS

> And I never fail to be surprised by the gift of
> an odd remainder, footloose at the end.
>
> —Mary Cornish

The giant step in computation that occurred during World War II derived in large measure from number theory, one of those esoteric areas of mathematics that seemed until recent times to be of interest only to ivory tower university theorists and a few amateur mathematicians. Despite the contributions of that subject to modern computing, this mathematical subdivision is still considered by many in the mathematics community to be out of the mainstream of contemporary research.

Although their activities take them far afield, the basic concerns of number theorists lie in the properties of the integers; that is, zero and the positive and negative whole numbers. This set is often displayed as $\{\ldots, -3, -2, -1, 0, 1, 2, 3, \ldots\}$, those terminal dots (ellipses) indicating that the numbers continue on indefinitely following the pattern of those listed in both positive and negative directions.

The integers are often displayed in elementary school mathematics classrooms on what has come to be called a *number line* (Figure 2.1)

Your first thought may well be: What could be easier than dealing with the integers? I promise you that almost anything could be. You need only turn to a number theory textbook to see that matters are not as simple as they may appear. Beyond the first few pages, the math rapidly gets out of hand. And some problems have been so difficult that they

Inside Your Calculator: From Simple Programs to Significant Insights   By Gerald R. Rising
Copyright © 2007 John Wiley & Sons, Inc.

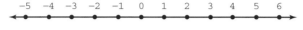

Figure 2.1 A number line.

have baffled mathematicians for centuries. Among them is the famous (or infamous) Fermat's last theorem. See Appendix D for more about this famous conjecture and its recent proof.

One reason why number theory is considered out of the mainstream of mathematics is that it supposedly provides so few applications to the real world; yet one of its most elementary topics, binary numeration, has given us the basis for what many consider a paradigm shift in computation. We will explore some aspects of that subject shortly.

TWO PROBLEMS

Consider now two problems that few calculator users consider seriously enough. Their solution will provide you with not just insight into procedures but also concepts that will prove useful later.

The First Problem

The first relates to how we enter numbers in our calculator. While this seems like a straightforward process, you will see that it has interesting implications.

Suppose that we wish to enter the number 342 in a calculator or computer. Everyone knows how to do that. You simply punch the keys 3, 4, and 2 in succession. No problem.

Well, not quite. How does the calculator know that the 3 you first pressed represents 300 and not just 3 or 30 or even 3000? Although we all write numbers from left to right, their values are determined from right to left. For the number 342, for example, we start on the right to assign 2 units, then 4 tens, and finally 3 hundreds. (If you fail to see this point for such a small number, consider how you would determine the value the 5 contributes to the number 5268433207. If you don't start counting places from the right, you have a system of which I am unaware.[1]

What we need is an algorithm that assigns a value to the number as we press those keys representing the digits. It turns out that the algorithm employed to carry this out is, once we think about it, transparently simple. We need only consider the process step by step in order to develop that procedure.

---

[1] The answer is, of course, 5 billion; or in Great Britain, 5 thousand million. (In Great Britain a billion is a million million.)

Step 1. You press the 3 key. At this point the calculator value is 3 and, if you next press a + or / or some other operation key, your calculator will know that 3 was the number you wished represented internally.

Step 2. In this case, however, it is not. The next key you press is 4, another digit, so the calculator must change its mind. Now the number it must consider is 34. How did it arrive at that? It shifted the 3 to the left and appended the 4. Again, if you next pressed an operation key, the calculator would operate on the resulting number, 34.

Step 3. But again you are not ready to calculate. You press the 2 key. And now the 34 is shifted to the left to make room for the 2, and we have 342.

That phrase "shift to the left" is exactly like the shift when we multiply by a number with more than one digit. (When we multiply by 23, for example, the partial product is also "shifted to the left" when we multiply by the 2.) The trick in both cases is the same; it is accomplished by multiplication by 10. In our input example, it is multiplication by 10 that changes 3 into 30 in step 2 and 34 into 340 in step 3.

We can generalize how those steps work in order to turn them into an algorithm:

Step 1. A number key is pressed (in our example the number 3.)

Step 2. If the next key pressed is also a digit, the earlier number is multiplied by 10 and the new number is added. (Pressing the 4 key gives us $3 \times 10 + 4 = 34$.)

Step 3. Repeat step 2 until a nondigit key is pressed; then continue with further calculation.

Another way of considering this is in the form of an algorithm that the calculator follows. To ensure that you see what is going on here, I have numbered the algorithm lines and added following the listing an explanation of how the algorithm works. One additional point: the $\rightarrow$ represents "is assigned to" or on your calculator $\boxed{\text{STO>}}$:

```
1    Press a digit key
2    Key pressed → N
3    Press another key
4    While the last key pressed is a digit
5         new key → X
6         10N + X → N
7         press another key
```

8   `End` of `While` loop until nondigit key is pressed

9   Continue calculations with N as defined

To see how this algorithm works, consider again entering our number, 342. Following the instruction in line 1, you press the 3, which is then assigned in line 2 to N. In line 3 you then press the next key, 4. You enter a loop consisting of lines 4–8, the `End` instruction in line 8 sending you back to line 4 until the condition in line 4 no longer holds. When that happens, you leave the loop and move to line 9.

Rather than continue this explanation in paragraph form, here in great detail are all of the steps required to enter 342 in your calculator:

Step	Line	What Happens
1	1	Press the first key, 3
2	2	3 is assigned to N
3	3	Press the next key, 4
4	4	The 4 is a digit, so we enter the loop
5	5	4 is assigned to X
6	6	Using the current values, $N = 3$ and $X = 4$, calculate $10N + X$, thus $10 * 3 + 4 = 34$ is assigned to N, replacing the former value
7	7	Press the next key, 2
8	8	`End` sends us back to step 4
9	4	The 2 is a digit, so we reenter the loop
10	5	2 is assigned to X
11	6	Using the current values, $N = 34$ and $X = 2$, calculate $10N + X$, thus $10 * 34 + 2 = 342$ is assigned to N, replacing the former value
12	7	Press the next key (any nondigit)
13	8	`End` sends us back to step 4
14	4	The new key is not a digit, so we leave the loop and go to step 9
15	9	Continue calculating with $N = 342$ (although we have no other use for it in this example, the calculator retains $X = 2$ as well.)

It is important to note that in carrying out this process the only actions the calculator user takes is pressing in succession the 3, 4, 2, and then any nondigit key. By following its internal algorithm, the calculator does the rest.

It is also worth noting here that only two internal storage places are required for this algorithm: storage for N and X. An alternate way of handling this problem would be to store each digit as its key is depressed and to make up the complete number when a nondigit key is punched. Clearly that would use more storage, especially with larger numbers. Our little algorithm will take care of decimal integers up to the usual 8- or 10-digit display capacity of the calculator.[2]

Here it is useful to compare the two ways of thinking about numbers algebraically. In standard school terminology we think of 342 as $3 * 100 + 4 * 10 + 2$. Following our algorithm, however, this number is represented differently. It becomes successively

$$3$$

$$3 * 10 + 4$$

$$(3 * 10 + 4) * 10 + 2$$

and we have $342 = (3 * 10 + 4) * 10 + 2$. For larger numbers, this new processing uses many parentheses. For example

$$56,832 = (((5 * 10 + 6) * 10 + 8) * 10 + 3) * 10 + 2$$

This is slightly shortened if we recall the alternate representation, $a * b = (a)b$:

$$56,832 = (((5 * 10 + 6)10 + 8)10 + 3)10 + 2$$

There is an attractive way of carrying out this process called *synthetic substitution*. Some readers will have met this procedure in school but may still not recognize this application. In the case of our last example, 56,832, the processing would look like the following.

We first write 10 representing the decimal base, and separately the digits in our number, 56832. Below this we leave space and draw a horizontal line:

$$10\ )\quad 5\quad 6\quad 8\quad 3\quad 2$$

$$\overline{\hspace{3cm}}$$

Now we start the process by simply bringing down the first digit, in this case 5:

$$10\ )\quad 5\quad 6\quad 8\quad 3\quad 2$$

$$\overline{\hspace{3cm}}$$

$$5$$

---

[2]This algorithm is designed to handle the input of integers only. It must, of course, be extended to handle decimals like 35.47.

Next we multiply the 5 in the lowest row by the 10 base and record the product (50) under the 6:

$$
\begin{array}{r|ccccc}
10\,) & 5 & 6 & 8 & 3 & 2 \\
& & 50 \\
\hline
& 5
\end{array}
$$

Add this second column and place the sum (56) below it. Notice that this multiplication and addition corresponds to our algorithm step of $5 * 10 + 6$:

$$
\begin{array}{r|ccccc}
10\,) & 5 & 6 & 8 & 3 & 2 \\
& & 50 \\
\hline
& 5 & 56
\end{array}
$$

Now repeat these two steps—multiply by 10 up to the right, then sum down. If you follow the algorithm with care, your final result should look like this

$$
\begin{array}{r|ccccc}
10\,) & 5 & 6 & 8 & 3 & 2 \\
& & 50 & 560 & 5680 & 56830 \\
\hline
& 5 & 56 & 568 & 5683 & 56832
\end{array}
$$

and the number you seek is at the end of the process. Note that this procedure avoids writing in all those parentheses.

We have seen that decimal integers can be written in two forms. It is useful to note that algebraic expressions called *polynomials* may similarly be written in these two forms. Thus, for example

$$9x^4 + 2x^3 - 8x^2 + 7x - 4 = (((9x + 2)x - 8)x + 7)x - 4$$

You can check this by multiplying out those parentheses of the right side one at a time, of course working from the inside out:

$$(((9x + 2)x - 8)x + 7)x - 4$$
$$((9x^2 + 2x - 8)x + 7)x - 4$$
$$(9x^3 + 2x^2 - 8x + 7)x - 4$$
$$9x^4 + 2x^3 - 8x^2 + 7x - 4$$

This, too, may be developed by synthetic substitution, but with x replacing the 10 of our decimal example, we have

$$
\begin{array}{c|ccccc}
x\,) & 9 & 2 & -8 & 7 & -4 \\
& & 9x & 9x^2+2 & 9x^3+2x^2-8x & 9x^4+2x^3-8x^2+7x \\
\hline
& 9 & 9x+2 & 9x^2+2x-8 & 9x^3+2x^2-8x+7 & 9x^4+2x^3-8x^2+7x-4
\end{array}
$$

You have seen here how the apparently simple act of entering a number in your calculator leads to some unexpected mathematical processing. You will soon see that this number processing will be useful in a different context.

THE SECOND PROBLEM

This side trip will take you back to elementary school, where you first took up the topic of division.

If you think back to that time, hopefully you will recall that, after you studied division exercises that "came out even" like 14/7 and 35/5, you found that some exercises left remainders (R) as in:

$$
\begin{array}{cc}
\quad 6 \ \text{R} \ 1 \qquad\text{and}\qquad \quad 7 \ \text{R} \ 8 \\
2)\,\overline{13} \qquad\qquad\qquad 10)\,\overline{78}
\end{array}
$$

or, if written in "long division" form:

$$
\begin{array}{cc}
\quad 6 \qquad\text{and}\qquad \quad 7 \\
2)\,\overline{13} \qquad\qquad\qquad 10)\,\overline{78} \\
\quad \underline{12} \qquad\qquad\qquad\quad \underline{70} \\
\quad 1 \ \text{R} \qquad\qquad\qquad\quad 8 \ \text{R}
\end{array}
$$

The introduction of fractions and, later, decimals soon replaced those calculations, and I suspect that many readers will have forgotten them. Asked to calculate 13 divided by 2 or on your calculator 13/2, you would answer $6\frac{1}{2}$ or 6.5, and asked to calculate 78/10, you would answer $7\frac{4}{5}$ or 7.8. To answer 6 remainder 1 and 7 remainder 8 to those requests now seems silly, childish, or even misleading.

For many programming applications, however, those calculations are far from silly. In fact, as you will see later, they will prove extremely useful to us in a number of settings.

It turns out that it is a bit complicated to calculate those integer quotients and remainders. If you try, for example, to divide 13 by 2 with the calculator, you obtain the expected decimal quotient 6.5 and certainly not

6 remainder 1. You need to do something quite different to obtain that answer.

There is a mathematical function that helps us with this situation. Mathematicians call it the *greatest integer function* (short for the "greatest integer less than or equal to" function) but computer scientists call it more simply "floor" and many calculators call it int. In elementary school terms you can best think of this function as "rounding down." For the positive numbers we are dealing with here, it is enough to think of this as simply dropping the fractional part. For negative numbers, however, rounding down doesn't work that way; thus $int(\pi) = 3$, but $int(-\pi) = -4$. In what follows, we will use int (short for the integer part) to represent this function.[3]

If you want the whole-number quotient of 13 divided by 2, then, you need only enter int(13/2). Your calculator should then display 6. For int(78/10) it should display 7. More generally, if you want the whole-number quotient of a number N divided by D, you would enter $int(N/D)$.

But now, how do you get that remainder? When dividing by 2, a complicated way would be to check to see if the division "comes out even" by means of a test like this:

```
If N/2 = int(N/2)
     Then 0→R
     Else 1→R
End
```

If $N = 12$, then $N/2$ would equal $int(N/2)$ because both would be 6. In this case the remainder R would be set equal to 0 by the Then instruction. But if $N = 13$, $N/2$ would equal 6.5 and $int(N/2)$ would equal 6. Since $6.5 \neq 6$, you would apply the Else instruction, which would give you a remainder of 1, as desired.

This approach has several limitations. It requires many instruction lines, control structures, and some complex processing. Far more important, it works only for division by 2. For other divisors, additional remainders are possible. For example, 78/10 produces a remainder of 8.

A better approach is to follow what happens when you actually carry out that "long division" process:

$$\begin{array}{r} 6 \\ 2\overline{)\,13} \\ \underline{12} \\ 1 \text{ R} \end{array} \qquad \text{and} \qquad \begin{array}{r} 7 \\ 10\overline{)\,78} \\ \underline{70} \\ 8 \text{ R} \end{array}$$

---

[3]Beware, however; you may have a calculator on which int simply drops the decimal part and gives a different answer for negative numbers. You would then need to work around this problem in your programming.

Consider first, 13/2. Recall that the quotient was obtained as int(13/2). To obtain the remainder, you multiply that quotient by 2, giving 12, and subtract that 12 from 13. Restated in terms of the quotient int(13/2) we calculated, that would be $13 - 2*\text{int}(13/2)$.

The same process applies to 78/10. The quotient is int(78/10), and the remainder is the dividend minus 10*int(78/10).

This process is perfectly general. For any positive integer N, the remainder, R, for N divided by D is $\text{R} = N - D*\text{int}(N/D)$. Pictured, the process looks like this:

$$
\begin{array}{r}
\text{int}(N/D) \\
\hline
D \ ) \ N \\
D * \text{int}(N/D) \\
\hline
N - D * \text{int}(N/D)
\end{array}
$$

Now we have the means of finding the quotient and remainder when dividing by D. Since int(N/D) = Q, we can use the two instructions:

int(N/D)→Q
N−D*Q→R

or, if we are interested in the remainder separately, we can write

N−D*int(N/D)→R

In carrying out serious mathematical programming you will find those instructions very useful and well worth remembering.[4] When you need them in this text, however, I will recall them for you.[5]

But now, having completed our detour and addressed our two problems, we will use what we have developed to show how decimal and binary numbers are related.

While this will provide examples for programs, it must be pointed out that calculators, unlike computers, do not calculate in binary. Instead they calculate using *binary-coded decimal* (BCD) number representation. In BCD only the digits of decimals are reproduced in binary. It will be clear in the chapters that follow, however, that these algorithms illustrate important tasks.

---

[4]In many computer languages and on more advanced calculators, you can obtain the remainder in the division of N/D by use of the modulo or mod function. For example, depending on your specific programming language, you would enter 78 mod 10 or mod(78,10) to obtain the remainder 8.

[5]In Appendix E you will find an application of integer division to long division and expressing fractions as repeating decimals.

BINARY NUMERATION

In base ten we have the digits less than ten, that is, 0 through 9. This Hindu-Arabic numeration format applies to any other numeration base as well. Base seven would utilize the seven digits less than seven: 0 through 6. Base twelve, the duodecimal system,[6] needs the digits less than twelve: 0 through 9 together with two more digits to make twelve. The additional digits for base twelve are usually written with the letters such as A and B. Base sixteen, a base used widely in computer science, requires six digits in addition to our ten to represent numbers. Thus you would count in base sixteen: 1,2,3,4,5,6,7,8,9,A,B,C,D,E,F,10,11, and so on.

It is worth generalizing these ideas to all base systems with base $N \geqslant 2$. Each system requires N digits, 0 through $N - 1$, but there is never any digit for N, the number base itself. In every one of these systems, the number N itself is written as 10.

Following this rule, base two (the binary system) utilizes only the two digits less than two: 0 and 1, and the number two is written 10.

Remarkably, with just this pair of digits we can still represent all the positive integers and, as we will see, by extension, all rational numbers, both positive and negative. We concern ourselves here, however, just with those positive integers and zero.

With virtually everything operating electronically today,[7] some useful working mechanisms are being lost to us. I offer one of these here: the automobile or motorcycle odometer. Until electronics took over, this was a mechanical device made up of rotating cylinders.

Those old odometers, a breakdown of which is displayed in Figure 2.2 operated on a simple principle. The rightmost dial was driven by a gear attached to one of the car or bike wheels. It turned as distance was covered. But this dial was attached to its neighbor to the left in such a way that when that first dial turned from 9 back to 0, it rotated the second dial one digit.

Suppose that you have turned back the dials to zeros so that your odometer display is 0000000.[8] As the vehicle moves forward, you will then display 0000001, 0000002, 0000003, up to 0000009, but then the "carry" will come into play as the rightmost dial turns from 9 to 0 and

---

[6]Base twelve occurs in English measurement—for example, 12 inches in a foot or a dozen and 12 dozen in a gross. Because the number 12 has {1,2,3,4,6,12} as factors while 10 has only {1,2,5,10} and for other reasons, the Dozenal Society (formerly the Duodecimal Society) continues to argue for wider use of base twelve. To learn more about the society and its work, visit its website: www.polar.sunynassau.edu/~dozenal/.

[7]One of the few devices that will probably not soon be taken over by electronics is the flush toilet of your home.

[8]Many odometers display tenths of a mile. We consider only integers here.

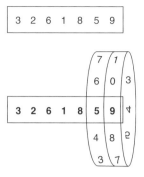

Figure 2.2   A decimal odometer.

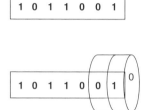

Figure 2.3   A binary odometer.

you next get 0000010. Similarly, after the display reaches 0000019, that same mechanism will produce 0000020.

Each dial except the one on the far left has this same carrying mechanism. When you reach 0000099, for example, carries on both the right-hand dials come into play and you get 0000100. I recall my brother and I, as youngsters, being excited whenever a group of 9s appeared and we could watch the transition to the next number with many of the dials rotating at the same time.

A binary odometer would work in the same way except that the number corresponding to the decimal 9 (the largest digit) is, since we have only 0 and 1, simply 1. Each of the odometer cylinders would have just two digits on them, and the hook to the next cylinder to the left would operate whenever a dial would change from 1 back to 0 (see Figure 2.3).

Let's see how this works. Again we'll turn our odometer back to zeros in order to begin with 0000000. As the rightmost dial turns, we'll first have 0000001 as before, but here we already need to have that rightmost cylinder change from 1 to 0. When it does, it pulls the next digit with it, and we have 0000010. Then 0000011, but this is like the decimal 0000099, and when those right two cylinders change to zeros, the next digit changes to 1. This produces 0000100.

Following this rule, here are the successive binary integers. I offer them with the odometer zeros and then without them, the way we more often think of numbers:

0000000	0
0000001	1
0000010	10
0000011	11
0000100	100
0000101	101
0000110	110
0000111	111
0001000	1000
0001001	1001 ...

Thus we would count in binary:[9] 1, 10, 11, 100, 101, ..., and these binary numbers correspond to our decimal 1, 2, 3, 4, 5, ... Here, then, are the decimal numbers from 1 to 20 with their corresponding binary numbers:

Decimal	Binary
1	1
2	10
3	11
4	100
5	101
6	110
7	111
8	1000
9	1001
10	1010
11	1011
12	1100
13	1101
14	1110
15	1111
16	10000

[9]It is not a good idea to assign the usual (decimal) number words, like ten for 10 and eleven for 11, to these binary numbers. Doing so can lead to confusion with decimal numeration. It is better to read binary numbers digit by digit; thus, 1101, for example, would be read "one one zero one" or more commonly "one one oh one." Thus you would count: "one, one oh, one one, one oh oh, one oh one," and so on.

17   10001
18   10010
19   10011
20   10100

If binary numeration is new to you, those integers probably seem weird. Even so, as you look at them you should immediately observe some important regularities. Notice first, for example, that all the binary integers with a 1 followed by all zeros correspond to the decimal integers 2, 4, 8, and 16, that is, powers of 2. Then notice that the binary integers represented by all 1s are always one less than powers of 2.[10] (Those are like the nines that change to zeros on the decimal odometer.)

We need to address one more matter before we consider how to change decimals to binary numbers and binary numbers to decimals. We should clarify the matter of notation, for it is very easy to confuse decimal and binary numbers if we are not careful. The binary 10 ("one oh") whose decimal value is 2, for example, looks exactly like the decimal 10 ("ten") whose value is five times greater. We need to avoid calling that binary 10 "ten." This problem can be met by adding a subscript after a number to indicate its base as in the aside above. Thus we could write: $20_{\text{decimal}} = 10100_{\text{binary}}$, and you will occasionally see this written as: $20_{10} = 10100_2$ or $20_{\text{ten}} = 10100_{\text{two}}$.

There is, of course, a problem with these latter forms because you must agree beforehand that the subscripts will be decimal numbers. In what follows I will use the last of these notations only when confusion between the bases might arise. More often the context will make clear with what base we are dealing.

PLACE VALUE AND CONVERSION

The great power of the Hindu-Arabic numeration system derives from the idea of place value. We met this idea earlier when we considered how numbers are entered into a calculator. It is important to understand that

---

[10]Clearly, the number of binary digits necessary to represent a number is larger than the number of decimal digits. In fact, for 8 and 9, four times as many binary digits are required, 1000 and 1001. Because we have five times as many digits for decimal numeration, you might think that—as a kind of tradeoff—you would often need five times as many digits in binary notation. That is not the case. In fact after 9, it is never necessary to use more than 3.5 times as many binary digits to represent a decimal. For those interested in the mathematics that supports this, Appendix O shows that this ratio of binary to decimal digits needed to represent large numbers approaches $1/\log 2 = 3.3219\ldots$.

this same place value system applies to binary numbers. Let's see how this works.

We know that when we write 3649, those digits have "place values":

Thousands	Hundreds	Tens	Units
3	6	4	9

Moving to the left digit by digit, each place carries a value 10 times the preceding place.

The same thing is true for numbers in binary, except that, as we proceed to the left the value of each digit is multiplied by 2. Thus, if we write 1101, we have:

Eights	Fours	Twos	Units
1	1	0	1

There is, of course, a problem with what I have written. Once again, I have used decimal names, this time for those columns. The word "two" is okay as that serves the same role in the binary system as the word "ten" does in the decimal system. It is the "four" and "eight" that cause the problem.

We can get around this in the following way. In decimal, we could have used exponents and evaluated 3649 as

Ten^3	Ten^2	Ten	Unit
3	6	4	9

and in binary we could have written

Two^3	Two^2	Two	Unit
1	1	0	1

The clear similarity between these two expressions of place value suggests that any Hindu-Arabic number may be written in *polynomial* form. If t is the base we are using, then the decimal number 3649 may be expanded as $3t^3 + 6t^2 + 4t + 9$, with $t = 10$, and the binary number 1101 may be expanded as $1t^3 + 1t^2 + 0t + 1$, with $t = 2$.

We have, of course, cheated a bit in these two expressions. We have again slipped into decimal mode in giving the value of t. This slippage, however, gives us one means of converting binary numbers to decimal. Simply substituting 2 for t (as in the song) in the polynomial, $1t^3 + 1t^2 +$

$0t + 1$, and calculating in decimal, we obtain successively:

$$1 * 2^3 + 1 * 2^2 + 0 * 2 + 1$$

$$1 * 8 + 1 * 4 + 0 + 1$$

$$13$$

and we have shown what we found in our odometer counting, that $1101_{two} = 13_{ten}$.

For larger numbers this process becomes a bit cumbersome, so we can instead deal with our polynomial by the left-to-right representation we introduced earlier in this chapter; thus $1t^3 + 1t^2 + 0t + 1$ becomes $((1t + 1)t + 0)t + 1$, which, evaluated with $t = 2$, gives us

$$((1 * 2 + 1) * 2 + 0)2 + 1$$

$$((3 * 2) * 2 + 1$$

$$6 * 2 + 1$$

$$13$$

This is more easily accomplished with the synthetic substitution format we also introduced for the evaluation of polynomials. To change 1101_{two} to decimal by this means, we simply write

$$2)\quad 1\quad 1\quad 0\quad 1$$

and forge ahead. Our result will look like this:

2)	1	1	0	1
		2	6	12
	1	3	6	13

For small numbers, either method is easy to use, but for numbers with many digits, synthetic substitution is far easier. Suppose, for example, that we have the number 101010101_{two}, which we wish to convert to decimal. The worked-out synthetic substitution would appear like this

2)	1	0	1	0	1	0	1	0	1
		2	4	10	20	42	84	170	340
	1	2	5	10	21	42	85	170	341

and we have established that $101010101_{two} = 341_{ten}$.

For the calculator to carry out this kind of mechanistic conversion, the process is straightforward. An algorithm like the one introduced to read numbers into the calculator may be used:

Leftmost digit $\rightarrow N$
While digits remain
     next digit $\rightarrow X$
     $2N + X \rightarrow N$
End of loop governed by the While test

Here are program steps that will carry out this algorithm:

```
PROGRAM:BINDEC
: 0→N
: Prompt B                 Leftmost binary digit
: While B=0 or B=1
:    2N+B→N
:    Prompt B              Next digit, or enter 2 to end loop¹¹
: End (While)
: Disp N
```

That is a very simple program, but it has one major drawback. To use it, you have to enter the binary digits one at a time. We would like to have a program that will allow you to enter at the prompt a binary integer like 1101 and have the program display 13 in response.

To solve our new problem we will not modify what we have developed, but will start over with a different approach. We'll now use the idea of place value and work out our answer by evaluating the binary digits beginning on the right.

To change that binary integer 1101 into decimal, we recall how we assigned place value to that number:

Two^3	Two^2	Two	Unit
1	1	0	1

We now simply assign the decimal values to the place values:

2^3	2^2	2	1
1	1	0	1

---

[11]This program does not follow the algorithm because you have no programming tool to test whether B is 0 or 1. The entry of 2 (or any numerical value not one or two) tricks the program into ending the loop. Although the algorithm's test is not available as a programming step, your calculator has it internally. Also the nonprogram comments appear on the right in regular text font to distinguish them from programming lines.

Having done that, we can easily add up our decimal values. Working from the right, we have

$$
\begin{aligned}
1 * 1 &&&=&& 1 \\
0 * 2 &&&=&& 0 \\
1 * 2^2 = 1 * 4 &&&=&& 4 \\
1 * 2^3 = 1 * 8 &&&=&& \underline{8} \\
&& \text{Sum} &=&& 13
\end{aligned}
$$

To accomplish all that, things get a little tricky. First, we need a way to pick off the digits one at a time.

To get the units digit, U, from $B = 1101$, we use the program line `B−10*int(B/10)→U`.

By now you should recognize what that instruction is doing. It is finding the remainder when B is divided by 10, by the process we introduced earlier in this chapter. And indeed, if you divide 1101 by 10, you get a remainder of 1.[12]

Now we want to move on to the next digit. In order to treat it as the units digit of a number, we need to get rid of that 1 we have already used and change 1101 to 110. Once again, we apply integer division: 110 is the integer quotient when you divide 1101 by 10. Thus our new value of B is obtained by `int(B/10)→B`.

We also need a mechanism to multiply the digits we obtain from each place by the appropriate power of 2. To do this, we set a value for the exponent, E, to 0 at the outset, since the units digit will be multiplied by $2^0 = 1$. Then we will increase this exponent by one as we evaluate each subsequent digit.

Here, then, is the program to accomplish all this:

```
PROGRAM:BINDEC2
: 0→D
: 0→E
: Prompt B
: While B>0
:    B−10*int(B/10)→U      Split off a digit.
:    U*2^E+D→D             Evaluate the digit.
:    int(B/10)→B           "Remove" that digit.
:    E+1→E                 Increase the exponent.
: End (While)
: Disp D
```

---

[12] Alert readers may be concerned here. We are applying decimal division to a binary number. Recall that in binary, however, the base two is written 10, just as in decimal the base ten is written 10. Despite this, you can follow the argument by thinking of the division as it would be calculated in decimal form.

We have now solved the first of our two problems. You can convert a binary number to a decimal number by means of this program. We are left with the reverse problem, converting a decimal number to binary. Although the process is not complicated once we justify it, that justification will take some effort.

We begin with a familiar example: convert 13_{ten} to binary. (You know the answer, of course, because you just saw how to convert in the other direction.) We can do this by working backward through the synthetic division we used to convert from binary to decimal. At first we know only the following:

$$2\,)\quad 1$$

$$\overline{}$$

$$1 \qquad\qquad\qquad 13$$

But the following facts are available to us:

1. All the digits in the top line will be 0s or 1s, because they will be binary digits.
2. All the numbers in the second line will be even, because they are obtained by multiplying the previous sum by 2.

The only way we could obtain that 13 then would be to have a 1 in the top line and 12 below it. That gives us

$$2\,)\quad 1 \qquad\qquad\qquad 1$$
$$\qquad\qquad\qquad\qquad\quad 12$$
$$\overline{}$$
$$1 \qquad\qquad\qquad 13$$

Now we ask ourselves how we obtained the 12. It was the product of 2 times the previous value in the bottom line. That entry must then have been 6.

$$2\,)\quad 1 \qquad\qquad\qquad 1$$
$$\qquad\qquad\qquad\qquad\quad 12$$
$$\overline{}$$
$$1 \qquad\qquad 6 \qquad 13$$

Now we repeat the process we used with 13. That 6 must be the sum of an even number plus 0 or 1. The only possibility for this is $0 + 6$, and we have:

$$2\,)\quad 1 \qquad\qquad 0 \qquad 1$$
$$\qquad\qquad\qquad\qquad 6 \qquad 12$$
$$\overline{}$$
$$1 \qquad\qquad 6 \qquad 13$$

Repeat this procedure in the next column, and you arrive at the binary result, 1101, in the top row.

2)	1	1	0	1
		2	6	12
	1	3	6	13

Unfortunately, that is a rather cumbersome process to follow to convert decimal to binary. But careful examination of the process shows two important facts:

1. Whenever the number in the lower line was odd, the digit in the top line was 1; whenever the number in the third line was even, the digit in the top line was 0.
2. As we progress to the left along that bottom line, the values are the whole-number quotients obtained by dividing the preceding numbers by 2. (the remainder was already used and is no longer considered.)[13]

We can combine these two facts into one process by noticing that fact 1 above is equivalent to

1'. The number in the top line is the remainder when the number in the bottom line is divided by 2.

It appears that our old friend—dividing by 2 to obtain quotient and remainder—is the key to this conversion, and indeed that is the case. In our example we divide 13 by 2, placing the quotient, 6, to our left and the remainder, 1, in our answer. Then we repeat this process with the 6, and so on.

To change the decimal number 13 to binary, then, we would go through the following steps

$$13/2 = 6 \text{ R } 1$$
$$6/2 = 3 \text{ R } 0$$
$$3/2 = 1 \text{ R } 1$$
$$1/2 = 0 \text{ R } 1$$

---

[13] You can check these points out with the previous and longer synthetic substitution example establishing the equivalence of $101010101_{two} = 341_{ten}$.

and we would take those remainders in reverse order to obtain the binary number 1101.

Happily, we have developed the procedure to obtain the quotient and remainder when dividing by 2. We can use the following two programming lines:

```
int(D/2)→Q
D−2Q→R
```

We must now address the second problem. It is one of representation. Remember that our procedure of successive divisions gave us the digits, but in reverse order. We certainly don't want to end up with the binary number with its digits reversed, so we need take into account this ordering.

We can respond to this problem just as we did in our program changing binary integers to decimal integers.

We're now ready for a full program to change a decimal number to binary:

```
PROGRAM:DECBIN
: 0→B
: 0→E
: Prompt D
: While D>0
:    D−2*int(D/2)→U        Splitting off a binary digit
:    U*10^E+B→B            Building the binary number
:    int(D/2)→D            Eliminating the decimal digit
:    E+1→E                 Increasing the exponent
: End (While)
: Disp B
```

Our two programs BINDEC2 and DECBIN allow us to change back and forth between binary and decimal integers. Although we do not need these programs for processing large numbers in calculators, we will find that the ideas that those programs illustrate will occur again and again in the chapters that follow.

I conclude this chapter with a brief comment about binary-coded decimals (BCDs), for that is the way numbers are entered and processed in calculators. In this system each decimal digit is individually translated to binary. Thus 3972015 would be represented as

$$0011 \quad 1001 \quad 0111 \quad 0010 \quad 0000 \quad 0001 \quad 0101$$

Notice how four binary digits are used for each decimal digit. Computers do not use this representation because space is saved using full binary representation. For example, seven-digit numbers in regular binary representation would require only 24 digits as opposed to the 28 digits of BCD.

There is more information about the electronics that support binary processing in Appendixes F and G, and some insight into how carries might be handled for BCD arithmetic may be interpreted from the discussion of multiplication of large numbers in Appendix L.

3

INTEGER POWERS

The most powerful force in the universe is compound interest.

—Albert Einstein

Integer powers[1] are to multiplication as multiplication is to addition. Just as we have for addition

$$b + b + b + b + b = 5b$$

we have for multiplication:

$$b * b * b * b * b = b^5$$

Thus it is quite reasonable, if we think of multiplication as multiple addition, to think of powers as a kind of multiple (multiple addition).

There are several ways to define integer powers. The most common one is the one we all learned in school:

$$b^n = 1 * b * b * \cdots * b \text{ for integer } n \geq 0$$
$$\backslash n \text{ factors } /$$

Your definition may not have had that 1 at the beginning, but it allows us to include $n = 0$ as well as integers $n > 0$ in our definition. Now the definition confirms that $b^5 = b * b * b * b * b$, but it also resolves the case where we have no factors, $b^0 = 1$.

[1] In common usage, the word *power* simply means strength. In mathematics *power* and *exponent* have distinct meanings. In the expression $b^e = p$, b is the base, e the exponent and p the power. Thus, for example, the powers of 5 include 5, 25, 125, 625, and so on.

Inside Your Calculator: From Simple Programs to Significant Insights   By Gerald R. Rising
Copyright © 2007 John Wiley & Sons, Inc.

An alternate definition has two parts:

$$b^0 = 1, b^{n+1} = b^n * b, \text{ for integer } n \geq 0$$

This is called a *recursive definition* because you must build (or recur) the values to reach the value you wish to determine.

Suppose, for example, that you wish to determine b^3 by this definition. There are two ways to do this. In the first, start with $b^0 = 1$. Now let $n = 0$ to give $b^1 = b^{0+1} = b^0 * b = 1 * b = b$. Next let $n = 1$ to give $b^2 = b^{1+1} = b^1 * b^1 = b * b$. Finally let $n = 2$ to give $b^3 = b^{2+1} = b^2 * b^1 = b * b * b$.

Without all that discussion, this method builds in recursive steps: $b^0 = 1, b^1 = b, b^2 = b * b, b^3 = b * b * b$.

The second way is to work backward. We want $b^3 = b^{2+1}$, so let $n = 2$ in the definition. The definition tells us that $b^{2+1} = b^2 * b$, but $b^2 = b^{1+1} = b * b^1$. Combining these gives us $b^3 = b * b * b^1$. Finally $b^1 = b^{0+1} = b^0 * b = 1 * b$; this last one is from the first part of the definition. Substituting this in $b^3 = b * b * b^1$ gives us $b^3 = b * b * b * 1$ or simply $b * b * b$.

Again, simplifying this procedure, the power breaks down as $b^3 = b^2 * b = b^1 * b * b = b * b * b$.

This may all seem like a great deal of effort to arrive at something that we knew to be much easier from our first definition. It turns out to be useful for us to think of powers in this recursive form to build a simple program to calculate them.

Here is a program for calculating B^E for positive integer E:

```
PROGRAM:POSIPOW[2]
: Prompt B,E
: 1→P
: For (I,1,E)
:    P*B→P
: End (For)
: Disp P
```

I consider that program a perfect gem. It is simplicity itself. Once the values of B and E are entered in the first step, the power P is set equal to 1 and you enter the For loop. Each time you pass through this For

---

[2]Naming programs, especially when the number of symbols allowed is restricted, is not always easy. Here I have named this program POSIPOW for *positive integral powers*. A real problem, however, is remembering what those abbreviations represented when you originally wrote them.

loop, the current value of P is multiplied by B. You are building the power step by step as in the recursive definition.

Thus we have

Pass through For loop	P
Before	1
1	B
2	$B*B$
3	$B*B*B$
4	$B*B*B*B$
....	

Notice that the program even works for $E = 0$. Why? Because, when $E = 0$, the For loop is not calculated. The value of E is already greater than the starting value, 1, of the loop. Thus P remains unchanged and we have $B^E = 1$.

It is worth turning aside for a moment to show that multiplication can also be defined recursively and calculated by a related program. A recursive definition of multiplication by a whole number is:

$$0*a = 0, \quad (n+1)*a = a + n*a \text{ for integer } a \geq 0$$

With this definition you calculate successively after $0*a = 0, 1*a = a, 2*a = a+a, 3*a = a+a+a$, and so on. The program that would accomplish this is

```
PROGRAM:INTMULT
: Prompt A,N
: 0→P
: For (I,1,N)
:    P+A→P
: End
: Disp P
```

This time, of course, P accumulates the product, not the power.

Returning to our program to calculate powers of B^E:

```
PROGRAM:POSIPOW
: Prompt B,E
: 1→P
: For (I,1,E)
:    P*B→P
: End
: Disp P
```

Simple though it is, this program raises an important issue, and the best way to expose that issue is to consider an historical example.

Peter Minuet is credited with purchase in 1626 of Manhattan Island from local Native Americans for trinkets valued at $24. How historians arrived at that particular value is beyond me, but we will proceed with that widely published payment. Many people think of this as a Dutch ripoff of uninformed aborigines. But suppose that those Indians invested their $24 (perhaps in real estate; it was too early for a casino) to provide an annual return of 8%, which they leave invested to be compounded annually. What would their $24 be worth in, say, 2010?

This is a standard investment problem. The principal, P, is multiplied by 1.08 each year to give a new P. The formula to carry this out is $A = P(1 + R)^N$, with R the annual rate of return and N the number of years. In this case we would have $A = 24(1.08)^{384}$.

We can calculate that 1.08^{384} with our power program, using $B = 1.08$ and $E = 384$. If you have entered that program, you should try this now, and once you have a value, compare that value and the time it took to calculate it with the value and time you get an answer by keying in 1.08 $\boxed{\wedge}$ 384 $\boxed{\text{ENTER}}$.

The mathematical results should be the same: 6.834741711E12 or 6,834,741,711,000.[3] But while you arrive at the same answer, the time it takes for your program to run is significantly longer than the time for the "direct" calculation by using your calculator's power key. What are those times? Using a stopwatch, on my calculator the program took 5.8 seconds, while the key gave an answer almost instantaneously.

Clearly our program is slow compared to the calculator's own processing for large powers, but it is still very fast. In this regard just think of multiplying $1.08 * 1.08 * 1.08 * \cdots$ with 384 factors even using a calculator. In the program that For loop is passed through 384 times during that 5.8 seconds; thus (not counting other steps in the program) it does so over 65 times per second. That's faster than your incandescent light

---

[3]Scientific notation gives us the means to represent numbers too large or too small for all of their digits to be included in the calculator display. The E12 at the end of this number translates into $*10^{12}$. Recalling that each multiplication by 10 moves the digit one place (to the right when positive, to the left when negative), this means that the decimal point in 6.834741711 must be moved to the right 12 places. Once we get beyond the last digit displayed, 1 in this case, we affix zeros. The zeros are not meant to be correct (e.g., a more powerful computer would show that this number would continue 384); rather, they indicate the magnitude of the number. When the scientific notation exponent is negative as in 3.75E$^-$4, we have $3.75 * 10^{-4}$. This time we move the decimal point to the left four digits, affixing zeros when necessary to give .000375. Scientific notation also gives scientists the ability to express how accurate a measurement is.

flashes each second.[4] Not bad, but even so not as quick as your calculator's hardwired program.

We can improve this speed in striking fashion. Here is a somewhat more complicated program to do this:

```
PROGRAM:QINTPOW
: Prompt B,E
: 1→P
: While E>0
:    E−2*int(E/2)→R
:    If R=1
:          P*B→P
:    B*B→B
: int(E/2)→E
: End
: Disp P
```

If you try that program, you will find that it calculates 1.08^{384} almost as fast as the hardwired calculator program.

To show you how that program works, I will take you on still another diversion—a diversion from a diversion in this case. I promise to return to tie things up by the end of the chapter.

There is an interesting method of multiplying called *duplation and mediation* that was probably used to multiply Roman numerals two millennia ago. The method requires only the ability to carry out three simple operations: multiply by 2, divide by 2, and add.

Suppose, for example, that we wish to multiply $23 * 35$. We write the factors at the head of two columns:

$$23 \quad 35$$

Next we divide the left column by 2, discarding any remainder, and multiply the right column by 2. We will then have

$$23 \quad 35$$
$$11 \quad 70$$

We continue in this way until the left column reaches 1. We will then have

$$23 \quad 35$$
$$11 \quad 70$$
$$5 \quad 140$$
$$2 \quad 280$$
$$1 \quad 560$$

[4]Most incandescent lights work at 120 hertz (Hz), which means 120 changes per second including both on and off; thus the light turns on and then off 60 times per second.

Now we do something strange. We cross out any number in the right column opposite an even number in the left. In this case we would cross out the 280:

23	35
11	70
5	140
2	~~280~~
1	560

Finally we simply add the numbers remaining in the right column to obtain our product:

23	35
11	70
5	140
2	~~280~~
1	560
	805

Notice that, although the numbers are quite different, this method also works when we interchange the factors:

35	23
17	46
8	~~92~~
4	~~184~~
2	~~368~~
1	736
	805

More even number partners crossed out this time, but the same product.

If you are especially alert, you will notice that this process is similar to the one we used in Chapter 2 to convert a decimal number to binary. We are, in fact, doing just that: converting the number in the left column to binary.

To see what is going on here, consider an easy question: Which of the following numbers would convert to a binary number with a one in its units column: 3456 or 3457? Of course, the answer to that question is 3457. Because it is odd, it has a binary digit one in its units column. The rest of the columns represent multiples of 2. Even numbers would have a zero in that units digit. Now let's see how that applies to $23 * 35$. We began by writing 23 and 35 in two columns

23   35

and that $23 = 2 * 11 + 1$. We take care of that remainder of 1 (or binary units digit) by multiplying it by 35. In effect, we have

$$23 \quad 1 \quad * \quad 35$$

Now we have 11 twos left after dividing by 2 and "discarding" that remainder of 1. (You have seen that we weren't really discarding it. In fact, we used it as a multiplier.) The number 11 is odd again, so we have a 1 in the twos column of the binary representation of 23. Thus we have so far

$$
\begin{array}{rcl}
23 & 1 & * & 35 \\
11 & 2 & * & 35
\end{array}
$$

and once that two is taken care of, we have 5 fours left. Odd again, so we have a one in the fours column of our binary representation:

$$
\begin{array}{rcl}
23 & 1 & * & 35 \\
11 & 2 & * & 35 \\
5 & 4 & * & 35
\end{array}
$$

But this time when we divide that 5 by 2, we have 2 in the eights binary column. Dividing 2 by 2 leaves a remainder of 0. This means that there are no 8s, so we do not count 8 * 35. We then have:

$$
\begin{array}{rcl}
23 & 1 & * & 35 \\
11 & 2 & * & 35 \\
5 & 4 & * & 35 \\
2 & 8 & * & 35
\end{array}
$$

Our last division gives us 1 in the 16 binary column, so we have

$$
\begin{array}{rcl}
23 & 1 & * & 35 \\
11 & 2 & * & 35 \\
5 & 4 & * & 35 \\
2 & 8 & * & 35 \\
1 & 16 & * & 35
\end{array}
$$

Adding up those values in the right column, we have $(1 * 35) + (2 * 35) + (4 * 35) + (16 * 35)$. Factoring out the 35 (or applying the distributive law) we have, as expected, $(1 + 2 + 4 + 16) * 35 = 23 * 35$.

Now, how does this apply to our Peter Minuet problem? (Notice that we are finally working our way back to our powers program.)

Consider what we seek in that problem: 1.08^{384}. With our first program we calculated that as

$$1.08 * 1.08 * 1.08 * 1.08 \cdots (384 \text{ factors})$$

but we have rules for powers that allow us to bunch the factors. Specifically, we know that $b^x * b^y = b^{x+y}$.

What we can do is break the 384 down into a sum of binary powers, just as we did in duplation and mediation by peeling off those odd remainders. Here we go:

Number	R	Power of 2
384	0	* 1
192	0	* 2
96	0	* 4
48	0	* 8
24	0	* 16
12	0	* 32
6	0	* 64
3	1	* 128
1	1	* 256

You can then write 384 as $128 + 256$ or $2^7 + 2^8$. You can also, of course, read the binary representation for 384 by taking the 1s and 0s from the bottom of the column upward: $384_{\text{ten}} = 110000000_{\text{two}}$.

For our purposes, using our rule for powers, we have $1.08^{384} = 1.08^{128} * 1.08^{256}$.

In this case, even though you have so few factors in the final product, you might think that you still have to calculate 1.08^2 as $1.08 * 1.08$ and 1.08^4 as $1.08 * 1.08 * 1.08 * 1.08$, and so on.

Ah, but notice something else about those powers. Once we have calculated 1.08^2, we can simply multiply it by itself—$1.08^2 * 1.08^2 = 1.08^4$. We can similarly multiply $1.08^4 * 1.08^4 = 1.08^8$, in the process cutting down still more on the number of multiplications.

Our digressions reaching an end, it is finally time to return to the program designed to calculate powers faster to see how it carries out this process. I have numbered the program lines to help refer to them:

```
   PROGRAM:QINTPOW
1  : Prompt B,E
2  : 1→P
3  : While E>0
4  :    E−2*int(E/2)→R
5  :    If R=1
```

```
6   :            P*B→P
7   :       B*B→B
8   :       int(E/2)→E
9   : End
10  : Disp P
```

If you run this program entering 1.08 for B and 384 for E, you can follow what happens though a few of the While loops.

But first you need two reminders from Chapter 2—another digression but at least a brief one. You may have forgotten, but the fourth program line calculates the remainder when E is divided by 2. Also the eighth line calculates the quotient when E is divided by 2. Once you recognize those two calculations, the rest is reasonably straightforward.

In the program P accumulates the power that will be reported at the end of the program. It is set equal to 1 in line 2. Now we will explore what happens in each pass through the While loop.

In line 4 we divide the exponent E by 2 to obtain the remainder. That tells us if we have a 1 in its units digit. In this case the remainder of 384/2 is 0, so the answer to the If test is false and we skip line 6. We do, however, square the base B in line 7 and divide our exponent E by 2 in line 8, before we go back to take our second pass through the loop. We leave the loop with $B = 1.08^2$ and $E = 192$.

The next time through the loop, the remainder for 192/2 is again 0, so again we skip the If instruction. This time we end the loop with $B = 1.08^4$ and $E = 96$.

It turns out in this processing that the If test will not come out to be "true" until the eighth pass through the loop when $B = 1.08^{128}$ and $E = 3$. Then finally the If test is passed and P becomes $1 * 1.08^{128}$. It is true on the last loop pass as well to make $P = 1 * 1.08^{128} * 1.08^{256} = 1.08^{384}$, which in this case is our final answer.

You should see in that series of values how the values of $(1.08)^N$ are multiplied by the current value of the power, P, whenever the Nth binary digit of 384 (from the right) is one. That's the number in the Rth column from the right, the remainder when the current exponent E is divided by 2. The number 384 has lots of 0s in its binary representation, so only a few of those powers are included in forming the product—specifically 1.08^{128} and 1.08^{256}.

Now let us estimate how much the new program has saved in processing. We know that the original calculation of 1.08^{384} required 384 multiplications, the first $1 * 1.08$. How many does the new program require?

There are two kinds of products calculated in the new program. First there are the products for B. Beginning with 1.08, the program had to calculate 1.08^2, 1.08^4, 1.08^8, 1.08^{16}, 1.08^{32}, 1.08^{64}, 1.08^{128}, and 1.08^{256}. But each of those products required only multiplying the previous value by itself. Thus, for example, $1.08^{32} = 1.08^{16} * 1.08^{16}$. Therefore, we need carry out only eight of those multiplications.

We also have to multiply to obtain the successive values of P. Those multiplications occurred when the value of R was 1. Recall that R was the remainder when the remaining exponent was divided by 2 or equivalently the current binary digit. For numbers of this size, having a 1 each time through would have meant nine such multiplications.

Each time through the loop we also had to do two divisions (in lines 4 and 8) and one more multiplication (in line 4) and we made nine passes through the loop. Counting division as multiplication (by an inverse) we have 27 more multiplications to count.

Combining these, we have $8 + 9 + 27 = 44$ multiplications instead of 384, a considerable savings. We can return to those 5.8 seconds it took for the simple program. Roughly translated into time, the new program should take about 44/384*5.8 seconds or about a two-thirds of a second, fast enough to compare with the speed of the calculator's power ⌃ key.

This is only an example of the kind of efficiency that can be hardwired into a calculator or computer.

One final extension for integer powers remains. We have considered powers B^E, for $E \geq 0$. What about negative powers?

This is an easy problem to resolve. We need only recall that $B^{-N} = 1/B^N$. Thus, if the exponent is negative, we need only calculate the corresponding positive power and take the reciprocal at the end of the program. For example, to calculate 7^{-3}, we calculate 7^3 and then divide one by that value to arrive at $1/7^3$.

Here is how our simple power program could be modified to accomplish this:

```
PROGRAM:INTPOW
: Prompt B,E
: abs(E)→F
: 1→P
: For (I,1,F)
:    P*B→P
: End
: If E<0
:    1/P→P
: Disp P
```

The changes in this program are straightforward. In the second step F is the positive value of the exponent E (whether E was positive or negative). Then the `For` loop operates with this value producing, when it has completed its job, the value B^F. Now the `If` instruction kicks in. If the value of E is negative, we need $1/P = B^{-F} = B^E$. If $E \geq 0$, $B^E = B^F = P$ and we leave the value of P unchanged.

Similar additional steps would modify the shortcut program `QINTPOW`, and we now are able to calculate integer powers. In the next two chapters we will consider rational powers: in Chapter 5 square root (results of the 1/2 or .5 exponent) and in Chapter 6, powers like $23^{.37}$.

Only when we have completed those chapters will we finally have accomplished as much as that remarkable single $\boxed{\wedge}$ key on your calculator.

We should not leave this chapter without completing the exercise posed by Peter Minuit's dealings with the Manhattan Indians. We calculated 1.08^{384}, but we still had to multiply that value by the $24 originally invested. By 2010 the Native Americans would have accumulated $164,033,801,100,000, an amount that would give every American Indian over $35 million. Perhaps their original bargain wasn't so bad after all.

4

SQUARE ROOT

> Root, hog, or die.
>
> —Denis William Brogan

Before we explore algorithms for extracting square roots, I remind readers of some notation. We have a general sense that finding a square root "undoes" squaring: $4^2 = 16$ and so $\sqrt{16} = 4$. While both of those equations are correct, a problem arises. What if we square a negative number? For instance, $(-4)^2 = 16$ also, but we still have $\sqrt{16} = 4$. We don't get back that -4 that we started with.

It turns out that several things are involved here, including some history.

First, that radix symbol, the $\sqrt{}$, was employed before negative roots were considered and it continues to represent the positive square root.[1]

Mathematicians are happy with that usage, since that makes $\sqrt{}$ a function. Why? Because functions must produce single values. If we allowed $\sqrt{16}$ to represent ± 4 (as too many students believe), then $\sqrt{}$ would not be a function.

Thus we have two quite different situations:

1. There are two square roots of all positive real numbers. Thus *a* square root of 16 may be either 4 or -4.

---

[1]Incidentally, the bar across the top of the number plays the role of parentheses. Thus you will find in many older books the notation $\sqrt{}(a + b)$ where most of us would write $\sqrt{a + b}$. Notice that your calculator adopts the older notation.

Inside Your Calculator: From Simple Programs to Significant Insights   By Gerald R. Rising
Copyright © 2007 John Wiley & Sons, Inc.

2. There is only one $\sqrt{\ }$ of a positive real number. $\sqrt{16} = 4$. A number like $\sqrt{16}$ is often read "radical 16" or "root 16" to make this difference clear. It is also, less clearly, referred to as "*the* square root of 16."

But the problem goes a bit further. When we write n^2, we don't know whether that n is positive or negative or even zero, and unless we are very careful, this, too, raises an unexpected difficulty. Suppose, for example, that we have $n = -4$ and we consider $\sqrt{n^2}$. Carry out the steps: $(-4)^2 = 16$ and $\sqrt{16} = 4$. Once again we don't get back what we started with. For this reason we cannot write $\sqrt{n^2} = n$ unless we specify that n ≥ 0. In fact, if $n < 0$, $\sqrt{n^2} = -n$. In summary, we have

$$\sqrt{n^2} = \begin{cases} n & \text{if } n \geq 0 \\ -n & \text{if } n < 0 \end{cases} \tag{I}$$

A new symbol proves useful here. It is the symbol $|n|$ for what is called the *absolute value of n*. When we know the number that n represents, the idea is a simple one. Students often refer to it as "the number without the sign." Thus we have both $|5| = 5$ and $|-5| = 5$ as well as $|0| = 0$. The absolute value of a number is always non-negative, but we must not say that $|n^2| = $ n, unless we know that n is positive.

In fact, the formal definition of absolute value is

$$|n| = \begin{cases} n & \text{if } n \geq 0 \\ -n & \text{if } n < 0 \end{cases} \tag{II}$$

and it should be immediately apparent (by comparing I with II) that the definitions of $\sqrt{n^2}$ and $|n|$ are exactly the same. This gives us an easier way to represent $\sqrt{n^2}$:

$$\sqrt{n^2} = |n|$$

and this clears up the problem of $\sqrt{n^2}$ when we don't know whether n is positive or negative.[2]

Absolute value has an important role to play in mathematics. In particular, when we talk about the distance between two numbers, we want that distance to be nonnegative. For that reason, we let $|a - b|$ represent the

---

[2]That equation $\sqrt{n^2} = |n|$ works both ways, of course. You can, for example, determine the absolute value of n, when you have access to the $\boxed{\sqrt{\ }}$ key, by entering $\sqrt{n^2}$. In similar fashion, you could calculate $\sqrt{n^2}$ by entering abs(n).

distance between a and b. It makes no difference whether $a > b$ or $a < b$ or even $a = b$, that distance $|a - b|$ will never turn out to be negative.

With that behind us, we can turn to the way positive square roots are calculated.

Until at least the 1980s, one of the algorithms regularly taught in junior high school was for square root calculation. It was a procedure learned by reluctant students for hundreds of years[3] and, of course, promptly forgotten by most. It seemed a strange procedure and, like so many of the others students and teachers had to master, few knew of any justification for it whatsoever.

To demonstrate how arbitrary the algorithm is, I'll take you through the steps, applying them to an example. We will follow this algorithm to calculate $\sqrt{5715.36}$.

1. Separate the given number (formally the radicand) into pairs of digits working left and right from the decimal point. Place a decimal point where your answer will appear above the original decimal point.

$$\sqrt{\overline{57}\ \overline{15.36}}$$

2. Find the nearest square less than or equal to the leftmost pair.

$$49 \le 57$$

3. Write this number beneath the pair and its square root above it as the first digit in your answer.

$$\begin{array}{r} 7\ \ . \\ \sqrt{57\ 15.36} \\ \underline{49} \end{array}$$

4. Subtract your square from the pair as in the long-division algorithm.

$$\begin{array}{r} 7\ \ . \\ \sqrt{57\ 15.36} \\ \underline{49} \\ 8 \end{array}$$

5. Bring down the next pair of digits.

$$\begin{array}{r} 7\ \ . \\ \sqrt{57\ 15.36} \\ \underline{49} \\ 8\ 15 \end{array}$$

6. Multiply your partial answer by 20 and write the product to the left of the result you obtained in step 4.

$$\begin{array}{r} 7\ \ . \\ \sqrt{57\ 15.36} \\ \underline{49} \\ 140\ |\ \ 8\ 15 \\ \dfrac{815}{140} \approx 5 \text{ so} \end{array}$$

---

[3]Historians tell us that both square root calculation methods described in this chapter were used by Babylonians over 2000 years ago.

7. Divide your result in step 5 by your answer
   in step 6 and write the quotient as the next
   digit in your answer.

$$
\begin{array}{r}
7\ 5. \\
\sqrt{\,57\ 15.36\,} \\
49 \\
\hline
140\ |\ \ 8\ 15
\end{array}
$$

8. Add this digit to your answer in step 6 and
   then multiply the result by it, placing the
   product under your answer from step 5.

$$
\begin{array}{r}
7\ 5. \\
\sqrt{\,57\ 15.36\,} \\
49 \\
\hline
140\ |\ \ \ 8\ 15 \\
+5\ | \\
145*5|\ \ \underline{7\ 25}
\end{array}
$$

9. Subtract and continue from step 5.

$$
\begin{array}{r}
7\ 5. \\
\sqrt{\,57\ 15.36\,} \\
49 \\
\hline
140\ |\ \ \ 8\ 15 \\
+5\ | \\
145*5|\ \ \underline{7\ 25} \\
90
\end{array}
$$

10. The completed algorithm would look like
    this in slightly abbreviated form:

$$
\begin{array}{r}
7\ \ 5.\ 6 \\
\sqrt{\,57\ 15.36\,} \\
49 \\
\hline
140\ |\ \ \ 8\ 15 \\
145\ |\ \ \ \underline{7\ 25} \\
1500\ |\ \ \ 90\ 36 \\
1500\ |\ \ \ \underline{90\ 36}
\end{array}
$$

And indeed $\sqrt{5715.36} = 75.6$, as you can check using the $\boxed{\sqrt{}}$ key on your calculator or by squaring 75.6.[4]

Like other teachers at that time, I taught that algorithm as a memorized and essentially mindless process, and it wasn't until years later that I discovered why it works. Like the early Greek mathematicians, I will consider it first as a problem in geometry. You should agree that finding the square root of 5715.36 is equivalent to finding the side of a square with area 5715.36.

In other words, we are faced with the square of Figure 4.1.

We begin by finding a square that will fit inside this one as in Figure 4.2.

---

[4]Doubting that any student I taught would recall that square root algorithm, a few years ago at a school reunion I asked a few of them about it. Only one had any recollection. His response: "It had something to do with multiplying by 20, didn't it?" As, in fact, it did.

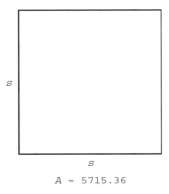

$A = 5715.36$

Figure 4.1

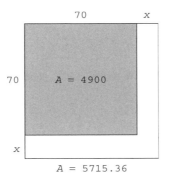

$A = 5715.36$

Figure 4.2

This corresponds to the first three steps in our algorithm, and we have

$$
\begin{array}{r}
7 \;\; 0. \;\; 0 \\
\sqrt{\;57\;15.36} \\
\underline{49\;00.00}
\end{array}
$$

or more simply

$$
\begin{array}{r}
7 \quad . \\
\sqrt{\;57\;15.36} \\
\underline{49}
\end{array}
$$

Clearly, if we subtract 4900 from 5715.36, we will have the area of that backward L-shaped blank part of the diagram.[5] Thus that remaining area is 815.36.

---

[5]L-shaped figures like that turned up so often in Greek mathematics that they were given a name, *gnomon*.

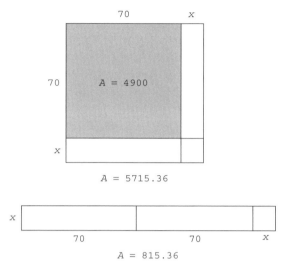

Figure 4.3

At this point the algorithm would show only

$$
\begin{array}{r}
7\quad. \\
\sqrt{57\ 15.36} \\
\underline{49} \\
8\ 15
\end{array}
$$

with that. 36 retained in the radicand until later.

If we let x be the rest of that side of the original square, that backward L-shaped area may be considered as two 70 by x rectangles and one x by x square. I have realigned those three quadrilaterals in Figure 4.3.

The resulting figure has dimensions $2 * 70 + x$ by x. Equivalently, we can think of that length as $20 * 7 + x$, and that $20 * 7$ is how we calculated the 140 in our algorithm:

$$
\begin{array}{r}
7 \\
\sqrt{57\ 15.36} \\
\underline{49} \\
140\ |\ 8\ 15
\end{array}
$$

The 140 is, of course, playing the role of first estimate of the length of that long narrow rectangle, and the x (which, you should recall, represents the rest of the side of the original square and thus the rest of our square root) is its width. We seek the value of x by dividing that 140 into the area 815.36; but clearly the quotient must be used in two ways, first to complete the length and then to serve as the width. When we decide to

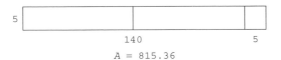

140   5
A = 815.36

Figure 4.4

choose $x = 5$, we have the following situation shown in Figure 4.4, which corresponds to that odd "add then multiply" step 8 of the algorithm:

$$
\begin{array}{r}
7 \quad 5. \\
\sqrt{57'\,15.36} \\
49 \\
\hline
\end{array}
$$

$$
\begin{array}{r}
140 \mid \quad 8\ 15 \\
145 \mid \quad \underline{7\ 25}
\end{array}
$$

Of course, even this does not take up the full area. We still have 90.36 left over:

$$
\begin{array}{r}
7 \quad 5. \\
\sqrt{57'\,15.36} \\
49 \\
\hline
\end{array}
$$

$$
\begin{array}{r}
140 \mid \quad 8\ 15 \\
145 \mid \quad \underline{7\ 25} \\
90\ 36
\end{array}
$$

We return to the original diagram to see how to proceed. We now have Figure 4.5.

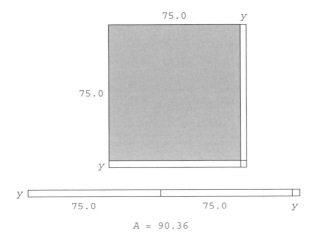

A = 90.36

Figure 4.5

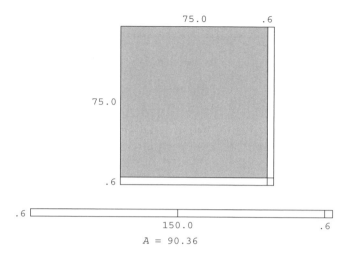

Figure 4.6

Here our trial divisor is 1500 (corresponding to 150.0 because we're working now with tenths), which is again 75 times our old friend 20. And y turns out to be 6, completing the procedure to give us the value of the square's side to be 75.6, as shown in Figure 4.6.

This corresponds to the completed algorithm:

$$
\begin{array}{r}
7 \quad 5. \ 6 \\
\sqrt{\ 57\ 15.36} \\
\underline{49} \\
140\ |\quad 8\ 15 \\
145\ |\quad \underline{7\ 25} \\
1500\ |\quad 90\ 36 \\
1506\ |\quad \underline{90\ 36} \\
0
\end{array}
$$

We can also approach this square root algorithm algebraically. To do so, I offer a simpler example: finding $\sqrt{676}$.

Let $\sqrt{676} = 10t + u$, with t the tens digit and u the units digit of the square root. Then, squaring each side and applying simple algebra, we have

$$676 = (10t + u)^2 = 100t^2 + 20tu + u^2 = 100t^2 + (20t + u)u$$

or, skipping those intermediate steps:

$$676 = 100t^2 + (20t + u)u$$

Notice in this representation how our friend 20 again plays a role. Clearly the largest value for the digit t we can use is 2, so we have

$$676 = 400 + (20 * 2 + u)u$$

which may also be written

$$676 - 400 = (20 * 2 + u)u.$$

This corresponds to

$$
\begin{array}{r}
2 \\
\sqrt{6 \quad 76.} \\
4 \\
\hline
40|\ 2\ \ 76
\end{array}
$$

Now all we need to do is choose a suitable u that satisfies $276 = (40 + u)u$, and that turns out to be 6. Adding that 6, multiplying by it, and placing it in our answer finally completes our algorithm since $276 = (40 + 6)6$:

$$
\begin{array}{r}
2 \quad 6 \\
\sqrt{6 \quad 76.} \\
4 \\
\hline
40|\ 2\ \ 76 \\
46|\ \underline{2\ \ 76}
\end{array}
$$

Needless to say, that algorithm—even when justified—was not a favorite of students.[6] Given that attitude, you can just imagine still earlier students who had to memorize an even more complicated algorithm for calculating cube root. That algorithm is based on the following relationship:

$$(10t + u)^3 = 1000t^3 + 300t^2u + 30tu^2 + u^3$$

$$= 1000t^3 + u(300t^2 + u(30t + u))$$

Never exposed to the geometric or algebraic justifications for that square root algorithm, it was popular with neither students nor teachers. That it was a rote and unjustified procedure was the reason often given for the introduction in schools of a replacement: an algorithm much simpler to state. It is often called the "divide and average" square root

---

[6]There is another aspect of it that caused student problems. In estimating the divisor after multiplying by 20, there is a tendency to overestimate, failing to account for the fact that you have to add before multiplying.

procedure. I know of no school text or classroom teacher who ever named the mathematician, Isaac Newton, whom historians credit for the method.[7] I will name it for him here.

NEWTON'S METHOD FOR SQUARE ROOT EXTRACTION

Here is the way Newton's method applies to the calculation of $\sqrt{N}$:

> Line 1.   Guess an answer. Call it G.
> Line 2.   Calculate $N/G = H$.
> Line 3.   If $G = H$, report G as $\sqrt{N}$.
> Line 4.   If not, average G and H to give a new G and go back to step 2.

Clearly, this is a much more straightforward procedure than the earlier algorithm. Let's see how it applies to our original example, $\sqrt{5715.36}$. (In doing so I will restrict computations to three decimal digits.)

Step 1, line 1.   We could guess any positive number we wish, but let's start as we did before with $G = 70$.

Step 2, line 2.   Calculate H:

$$\frac{5715.36}{70} = 81.648$$

Step 3, line 3. $G = 70$ and $H = 81.648$ are not equal, so we continue.

Step 4, line 4.   Average G and H:

$$\frac{81.648 + 70}{2} = 75.824$$

and 75.824 becomes a new G to return to line 2 of the algorithm.

Step 5, line 2.   Calculate a new H:

$$\frac{5715.36}{75.824} = 75.377$$

Step 6, line 3. $G = 75.824$ and $H = 75.377$ are not equal, so we continue.

---

[7]Although this method was used much earlier by the Babylonians, their use appears to be similar to that of twentieth-century school students, as a rote procedure. Newton provided the basis for the method as shown in Appendix I.

Step 7, line 4. For our next G, we average G and H:

$$\frac{75.824 + 75.377}{2} = 75.601$$

Step 8, line 2. Calculate a new H:

$$\frac{5715.36}{75.601} = 75.599$$

Step 9, line 3. $G = 75.601$ and $H = 75.599$ are close but still not equal, so we continue.

Step 10, line 4. Average G and H:

$$\frac{75.601 + 75.599}{2} = 75.6$$

Step 11, line 2. Calculate a new H:

$$\frac{5715.36}{75.6} = 75.6$$

Step 12, line 3. Since $G = 75.6$ and $H = 75.6$ are equal, we report 75.6 as $\sqrt{5715.36}$.

In summary, here are our successive values of G and H as we pass though the algorithm steps:

G	H
70	81.648
75.824	75.377
75.601	75.599
75.6	75.6

Newton's method has several positive features. It is easy to recall: "guess then (divide, average, repeat)." It also converges on the answer rapidly. And, remarkably, it is even error correcting; that is, if you make a computational error along the way, the procedure will right itself in subsequent steps.

At the same time this algorithm has features that make it a less than perfect procedure for paper–and–pencil computation. To get a sense of the first problem, you need only perform those steps, doing the long division by hand. Unless you are different from the rest of us who are unenthusiastic about carrying out long division, you will not be happy with this task.

There is another aspect of this problem that I have largely hidden in my recordkeeping. In the algorithm steps I rounded the numbers to three decimal places. Here is that list of successive Gs and Hs to 10-digit accuracy:

G	H
70	81.648
75.824	75.37666174
75.60033087	75.59966913
75.6	75.6

Our calculation did not need those extra digits, but for more accuracy they would be required, making the long-division process without a calculator even worse.

Before calculators became available, students hated that method even more than they disliked the earlier algorithm. I know: I had to teach it through the 1950s and 1960s. But then the early calculators appeared, and you would think that they would make this method more acceptable. Even with a calculator, however, recopying those values in order to carry out the calculations is both time-consuming and an invitation to recordkeeping errors.

With the earliest calculators (without any ☑ key, of course), there was a still easier approach. Seeking $x = \sqrt{N}$ is equivalent to seeking x with $x^2 = N$. Simply guess successive values of x and adjust them depending on whether their squares are larger or smaller than N. (Note that students didn't need a special key to square a number; they could merely multiply the number by itself.)

Here, for example, is how you might find $\sqrt{5715.36}$ by this means. We seek an x such that $x^2 = 5715.36$. As we did before, we'll start with a guess of 70.

Guess	Square	
70	4900	Too small; thus the answer is between 70 and 80
75	5625	Too small; thus the answer is between 75 and 80
78	6084	Too large; thus the answer is between 75 and 78
77	5929	Still too large; thus the answer is between 75 and 77
76	5776	Still too large; thus the answer is between 75 and 76
75.5	5700.25	Too small; thus the answer is between 75.5 and 76
75.7	5730.49	Too large; thus the answer is between 75.5 and 75.7
75.6	5715.36	Bingo!

Since $75.6^2 = 5715.36$, $\sqrt{5715.36} = 75.6$. If you try that method to calculate any other square root, you'll find it very straightforward and reasonably fast. And you can, of course, calculate by this means to any number of digits.[8]

There is a second problem with Newton's method that is still more important and must be addressed. When it was introduced in schools, and despite its sponsors' criticism of the earlier method for failing to justify their procedure, none of those sponsors justified this procedure any more than they did that first algorithm. We will address that justification after we have considered the calculator program.

It turns out that both of these last two procedures—Newton's method or guess and check by squaring—may be programmed, but we will choose Newton's method.[9]

Here is a simple program that will calculate $\sqrt{N}$ for any nonnegative number, N.[10]

```
PROGRAM:SQRT
: Prompt N
: Prompt G              Initial estimate G
: 0→H
: While G≠H
:     N/G→H             Divide N by your estimate, G, to obtain H
:     (G+H)/2→G         Average G and H to provide a new estimate, G
: End                   Go through the loop again until G = H.
: Disp G
```

If you run that program with a reasonable first estimate, G, you will find that it carries out its calculation in less than a second and about as fast as the programmed ☑ key on your calculator.[11]

To see what is going on, I will add two additional instructions and, since it is then superfluous, remove the final instruction:

---

[8]In 1969 my doctoral student, Paul Pang, now a college president, tested these three methods with middle school students. The clear winner in terms of both preference and understanding was this third method.

[9]An adaptation of the guess and check by squaring method turns out to be one of those used to solve complicated equations. It appears in Appendix M. As it happens, it, too, is a process credited to Isaac Newton.

[10]In the program the instruction `While G≠H` could be replaced with `Repeat until G=H`. With this instruction the check whether $G = H$ is at the end of the loop. Thus the loop is processed at least once. For this reason the step $0 \to H$ may be omitted.

[11]$\sqrt{N}$ may also be calculated by N ☐.5. As we will see in the next chapter, however, the programs may be very different.

```
PROGRAM:SQRTCK
: Prompt N
: Prompt G
: 0→H
: While G≠H
:     N/G→H
:     Disp G,H          Show the current value of G and H.
:     Pause
:     (G+H)/2→G
: End (While)
```

If you run that program with $N = 5715.36$ and $G = 70$, you will alternate the following values of G and H:

G	H
70	81.648
75.824	75.37666174
75.60033087	75.59966913
75.6	75.6

You should recognize those values as exactly the same ones we obtained by carrying out the algorithm earlier. That should certainly be the case because the algorithm simply replicates the steps for hand calculation.

TWO ASIDES ABOUT CALCULATOR CALCULATION

There is a problem with the While loop in the program we have used. That check whether $G \neq H$ can occasionally cause a calculator or computer to run endlessly. It may seem easy to check whether two numbers are equal, but very small differences between numbers that do not show up on your calculator display may make two numbers that appear the same yet turn out to be different. The result is the calculator values of G and H jumping back and forth between two nearly equal values. This is called *dithering*.

   To avoid this problem, your calculator probably does several things. First, as you have seen, it calculates with numbers more accurate than those shown in the display.

   Despite this additional accuracy, your calculator could still run into problems with minute differences between numbers—the rightmost unseen digit hovering between 4 and 5 and thus causing rounding differences, for example—and this can lead to the kind of thing that occasionally punishes us when our computer acts up and we have to press ESCAPE or even pull the plug.

To get around this, the calculator may check for minor differences. For example, it might apply the following rule: If it finds that the difference is less than half the value of the rightmost digit in your display, it will allow the numbers to be considered equal.

Programmers prefer to avoid this kind of error by providing this instruction themselves. For example, for our statement `While G≠H`, they might substitute something like `While abs(G−H)/G>.00000000005`.[12] A similar substitution can be used to test equality.

There is another problem with the program I have introduced. Your calculator $\boxed{\checkmark}$ key does not require you to enter an initial guess, G. It only calls for a value of N. To solve this problem, instead of asking for G in the program, we can use an arbitrary value of G, say, 10. Here then is the revised program:

```
PROGRAM:SQRTG10
: Prompt N
: 10→G                    Substituting for Prompt G
: 0→H
: While G≠H
:       N/G→H
:       (G+H)/2→G
: End
: Disp G
```

Let's see what that initial guess, clearly a poor one for this calculation, does to our computation of $\sqrt{5715.36}$. Here are the successive (but unreported[13]) values of G and H when this program is run:

G	H
10	571.536
290.768	19.6560832
155.2120416	36.82291619
96.0174789	59.52416233
77.77082061	73.48977361
75.63029711	75.56971503
75.60000607	75.59999393
75.6	75.6

---

[12]Technically $|G − H|/G$ is the *relative error* between H and G relative to G.
[13]We can obtain these values by running the program SQRTCK on page 64 with $N = 5715.36$ and $G = 10$.

Notice that, despite the wide swings of the first few values, the program zeroes in on the answer in only a few additional steps, and the calculator in large part makes up for these additional steps with its rapid processing.

The choice of 10 for G in the program was arbitrary and the calculation of the square roots of very large numbers, say, 10^{25}, or very small numbers, say, 10^{-25}, would take enough steps to make the time waiting for an answer evident in comparison to the almost instantaneous calculation using the ☑ key. Clearly, then, in this case the calculator algorithm is more sophisticated than ours.

ARITHMETIC AND GEOMETRIC MEANS

So far, you have been asked to accept Newton's algorithm without justification. To establish that it does indeed give us correct answers, we now provide the math that supports this process.

The "trick" of Newton's algorithm is to use the arithmetic mean as an approximation for the geometric mean in his algorithm. To understand his procedure, we need to define those means.

To find the average of two numbers, we calculate half their sum. Mathematicians assign the more formal name *arithmetic mean* (A.M.) to that average, so, for two numbers a and b, we have

$$\text{A.M.} = \frac{a+b}{2}$$

There are other types of means as well[14] and the one we are concerned with here is called the *geometric mean* (G.M.). Like the A.M., it, too, has an alternate name: the mean proportional.[15] For two positive numbers, it is defined as the square root of their product. Thus for a and b, we have

$$\text{G.M.} = \sqrt{ab}$$

[14]Among them are the harmonic mean:
$$\frac{1}{\frac{1}{2}\left(\frac{1}{a}+\frac{1}{b}\right)} = \frac{2ab}{a+b},$$
and the root mean square,
$$\sqrt{\frac{a^2+b^2}{2}}.$$

Mathematicians generalize all of these means to apply to more than two values a and b.
[15]The mean proportional between two numbers a and b is the number x in the proportion: $a/x = x/b$. To see that x corresponds to the G.M., solve the proportion for x: $x^2 = ab$ and $x = \sqrt{ab}$.

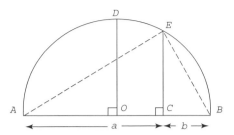

Figure 4.7 Arithmetic and geometric means.

We can compare the A.M. and G.M. for given a and b geometrically in Figure 4.7.

The given segments a and b are placed end to end as shown in Figure 4.7, with $a = AC$ and $b = CB$. The midpoint O of AB is located, and a semicircle is constructed with AB as diameter. Segments OD and CE are constructed perpendicular to AB, meeting the semicircle at D and E.

Since $a + b = AB$, the diameter of the semicircle, and OD is a radius of that semicircle, $OD = (a + b)/2$. Thus, OD is the A.M. of a and b.

An angle inscribed in a semicircle is a right angle, so angle AEB is a right angle. Thus the triangles ACE and ECB are similar, and we have the proportion $a/CE = CE/b$ so $(CE)^2 = ab$ and $CE = \sqrt{ab}$. We have shown that CE is the G.M. of a and b.

An important feature of this diagram is the comparison of OD and CE. Notice that varying the position of C on AB changes the relationship between a and b, but for any position, $OD \geq CE$. In fact, they will be equal only when $a = b$. This establishes that A.M. $\geq$ G.M.

We can show this useful relationship[16] algebraically as well. We begin with a statement that is always true for positive numbers, a and b; since any number squared is nonnegative: $(a - b)^2 \geq 0$. Squaring, we have $a^2 - 2ab + b^2 \geq 0$. Add $4ab$ to both sides of the inequality to obtain $a^2 + 2ab + b^2 \geq 4ab$. The left side is again a perfect square: $(a + b)^2 \geq 4ab$, and, since both sides are positive, we can take their square roots to obtain $a + b \geq 2\sqrt{ab}$. Finally, dividing by 2 gives us the inequality we seek, $(a + b)/2 \geq \sqrt{ab}$, which establishes once again that: A.M. $\geq$ G.M.

Now we have the tools we need in order to investigate Newton's algorithm for square root.

---

[16]For several interesting applications of this relationship, see P. J. Nahin, *When Least Is Best*, Princeton University Press, 2004, especially pages 13–20.

WHY NEWTON'S ALGORITHM FOR SQUARE ROOT WORKS

Newton's Method involves a series of what we have called *guesses*, which we have designated *G* in the program. Estimates may be a better word for them, but we will stick with *G*.

Look at that program again. I have numbered the steps for reference:

```
      PROGRAM:SQRT
1     : Prompt N
2     : 10→G
3     : 0→H
4     : While G≠H
5     :    N/G→H
6     :    (G+H)/2→G
7     : End
8     : Disp G
```

Now recall the roles of those three letters, N, G, and H. We seek $\sqrt{N}$ and the program calls for the value of N to process in line 1. Line 2 assigns that arbitrary value of 10 to G. In line 5, H plays the role of the quotient of the division step of "Divide and average." It is set at 0 in line 3 so that there is no possibility that it will be equal to that initial guess G.[17] This is necessary in order to enter the `While` loop.

Clearly the values of G and H change as the loop repeats. In order to examine these changes we need to designate them with different symbols, and this is a perfect role for subscripts. We set the initial values of G and H to be G_0 and H_0. Then subsequent values will be $G_1, G_2, G_3, \ldots$ and $H_1, H_2, H_3, \ldots$[18] and when we want to talk about a general value of G or H, we can designate that value as G_n or H_n.[19] Using these subscripts, we can describe how we process the algorithm from step to step as $H_n = N/G_n$ and

$$G_{n+1} = \frac{G_n + H_n}{2}.$$

[17]Calculator and computer users often forget that values left from previous programs continue to reside in memory. Here, for example, the concern is that a leftover $H = 10$ would lead to $\sqrt{N} = 10$ for any N. Any negative value for H would have served equally well in step 3. In this program with G assigned the value 10, any number different from 10 would do for H.

[18]Notice how these subscripts allow us to recall the basic variable while separating its values at stages in the process.

[19]Do not confuse this n for a general step number with the N for which we seek the square root.

If this algorithm is to work, it must provide a series of values that will get closer and closer to $\sqrt{N}$, finally getting so close that we will be able to discern no difference between that value and $\sqrt{N}$.[20] This suggests that we study the error at each stage of the process. That error, which we will call E_n because it differs at each step n, we define as $E_n = |G_n - \sqrt{N}|$.

Note several things about that definition. First, the error is changing (hopefully reducing) and that is why the subscript n is used. For each G_n, there is a corresponding E_n. The second thing about the definition is the use of absolute value, which was defined earlier in this chapter. The reason for its use here is our wish to have E_n positive. If $G_n \geq \sqrt{N}$, $G_n - \sqrt{N}$ would be positive (or zero), but if $G_n < \sqrt{N}$, $G_n - \sqrt{N}$ would be negative; in either case the absolute value would make the corresponding value of E_n positive.

Dealing with absolute value algebraically is rather complicated, however. For that reason it will be useful to show that, for values of $n \geq 1$, $G_n \geq \sqrt{N}$. In that case we can remove the absolute value signs because $G_n - \sqrt{N}$ will be nonnegative and the absolute value of a nonnegative number is that number.

Fortunately, no matter what value we choose for G_0, it is straightforward to show that successive values of G_n will be greater than $\sqrt{N}$. Here is how we can do this.

We have established that

$$H_n = \frac{N}{G_n} \qquad \text{and} \qquad G_{n+1} = \frac{G_n + H_n}{2}.$$

Note that the first equation is equivalent to $G_n H_n = N$ and therefore the geometric mean of these values, $\sqrt{G_n H_n} = \sqrt{N}$.

We also know that the A.M. $\geq$ G.M. Applying this to the variables G_n and H_n, we have

$$\frac{G_n + H_n}{2} \geq \sqrt{G_n H_n}$$

Substituting what we have established for each member of this equation, we have what we set out to show that $G_{n+1} \geq \sqrt{N}$ for $n = 0, 1, 2, 3, \ldots$.

Notice that this statement does not say that $G_0 \geq \sqrt{N}$. (In fact, our choice of $G_0 = 10$ in our program would be less than $\sqrt{N}$ for any $N \geq 100$.) But it does say that, no matter what the choice of a positive G_0, $G_n \geq \sqrt{N}$ for $n = 1, 2, 3, \ldots$. To see that this is true, simply

---

[20]In some cases as with $\sqrt{64}$ the value will be exact, but in others, like $\sqrt{5}$ when the root is irrational, the value will be as near as your calculator will display.

substitute the given values for n successively in the previous statement for G_{n+1}.

It is not necessary, but we could have set G_0 in our program so that it, too, would be greater than $\sqrt{N}$. In place of the statement 10→G, we could have written N+1/4→G. To show that $N + 1/4 \geq \sqrt{N}$, we simply substitute 2N for a and 1/2 for b in the A.M. $\geq$ G.M. theorem

$$\frac{a+b}{2} \geq \sqrt{ab}$$

The resulting statement,

$$\frac{2N + \frac{1}{2}}{2} \geq \sqrt{2N * \frac{1}{2}},$$

reduces to the desired $N + 1/4 \geq \sqrt{N}$.

Knowing that $G_n \geq \sqrt{N}$ at least for all $n \geq 1$, we consider how the corresponding H_n relates to $\sqrt{N}$. Taking reciprocals of inequalities relating positive values reverses the order, thus $G_n \geq \sqrt{N}$ leads to

$$\frac{1}{G_n} \leq \frac{1}{\sqrt{N}}$$

Multiplying each side of this inequality by N gives

$$\frac{N}{G_n} \leq \frac{N}{\sqrt{N}}$$

Since the left side is H_n and the right $\sqrt{N}$, we have shown that $H_n \leq \sqrt{N}$.

We now have the important relationship $H_n \leq \sqrt{N} \leq G_n$ for $n = 1, 2, 3, 4, \ldots$. In other words, we have $\sqrt{N}$ between H_n and G_n.

We're finally ready to explore those errors. Recall that we defined the error $E_n = |G_n - \sqrt{N}|$. Since we have proved that $G_n \geq \sqrt{N}$ for $n \geq 1$, however, we can drop the absolute value signs to give $E_n = G_n - \sqrt{N}$.

If we can show that the error is shrinking significantly from step to step, we will have proved our algorithm works. What we seek then is the relation between E_{n+1} and E_n. Specifically we want $E_{n+1} < E_n$ as we approach $\sqrt{N}$. We will begin with the result we just proved and build upon it. You should be able to follow these steps:

$$H_n \leq \sqrt{N}$$

$$G_n + H_n \leq G_n + \sqrt{N}$$

$$\frac{G_n + H_n}{2} \leq \frac{G_n + \sqrt{N}}{2}$$

$$G_{n+1} \leq \frac{G_n + \sqrt{N}}{2}$$

$$G_{n+1} - \sqrt{N} \leq \frac{G_n + \sqrt{N}}{2} - \sqrt{N}$$

$$G_{n+1} - \sqrt{N} \leq \frac{G_n + \sqrt{N} - 2\sqrt{N}}{2}$$

$$G_{n+1} - \sqrt{N} \leq \frac{G_n - \sqrt{N}}{2}$$

$$E_{n+1} \leq \frac{E_n}{2}$$

We have proved that the error is at least halved in each pass through the algorithm. If each $E_{n+1} = E_n/2$, we would have the situation depicted in Figure 4.8.

That we actually do much better than this we can see by following a simple example, the program calculation of $\sqrt{9}$. We expect $E_{n+1} \leq$

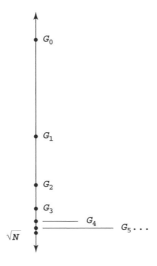

Figure 4.8 G_n values approach $\sqrt{N}$.

$E_n/2$, which is equivalent to expecting $E_{n+1}/E_n \leq \frac{1}{2}$. The fourth column calculates the successive values of E_{n+1}/E_n:

n	G_n	$E_n = G_n - \sqrt{N}$	E_{n+1}/E_n
1	9.25	6.25	
2	5.111486486	2.111486486	<0.34
3	3.436113369	0.436113369	<0.21
4	3.027675872	0.027675872	<0.07
5	3.000126492	0.000126492	<0.005
6	3.000000003	0.000000003	<0.00003
7	3	0	0

In this tabulation, note that the error not only decreases to less than half at each step but also decreases more rapidly as the algorithm progresses. You can see in Appendix H that, once you have a correct decimal fraction digit, in other words when you are within one of $\sqrt{N}$, the number of correct decimal digits at least doubles with each successive step.

You now have not only a fast algorithm that carries out the work of your calculator's $\sqrt{N}$ key but you also have an algorithm that you can use as part of the process for other keys. Thus $\sqrt{N}$ can play the role of what is called a *subroutine* in other calculations.

From now on, then, we need not write out those program steps. We will call on that subroutine merely by writing $\sqrt{N}$ in our programs. Whenever we do this, we will know that the calculator could be sending N into a modification of that nine-step program. As you will see in later chapters, this gives us an important tool for designing additional algorithms.

Several final notes. Readers familiar with the calculus can see in Appendix I how this "divide and average" algorithm is indeed a particular case of the more general Newton's Method. Appendixes J and K provide some ways to approximate irrational numbers with increasingly accurate fractions. Appendix J also shows how the Greeks proved the existence of an irrational ratio.

5

RATIONAL POWERS

> To him who looks on the world rationally, the world in its
> turn presents a rational aspect. The relation is mutual.
>
> —Georg Wilhelm Friedrich Hegel

RATIONAL AND REAL NUMBERS

At the outset I seek to clarify something about rational and real numbers. Most readers who have studied mathematics in school or college know that the rational numbers include the integers and fractions. They may be defined as all numbers that can be written as a/b, with a any integer and b any positive integer.[1] Thus, numbers like $\frac{7}{283}$ and $-\frac{59}{13}$ are rational numbers. The integers fit this definition as well, since they may be written with 1 in the denominator. For example, $-37 = -\frac{37}{1}$. So too are decimal fractions like $\frac{379}{1000} = .379$ or $-\frac{21}{10} = 2.1$ rational numbers.

But there are also numbers that are not among the rationals. They are called *irrational numbers*. These include numbers like $\sqrt{7}$, π, e, and $\cos 27°$. The rational and irrational numbers together make up the real numbers.[2]

---

[1]They may also be defined as numbers that may be expressed as repeating decimals. Thus numbers such as $\frac{1}{3} = .33333\ldots$, and $\frac{3}{7} = .428571\,428571\,42857\ldots$ are rationals. Repeating zeros are also allowed in order to make numbers that do not seem to need repeating to be decimals, like 27 and $\frac{1}{2}$, for example, fit this category: $27 = 27.00000\ldots$, and $\frac{1}{2} = .500000\ldots$.

[2]Unlike rational numbers, irrational numbers cannot be expressed as repeating decimals.

Inside Your Calculator: From Simple Programs to Significant Insights   By Gerald R. Rising
Copyright © 2007 John Wiley & Sons, Inc.

Now here is where some confusion arises. Your calculator has a key marked π, and we have a calculator key (and a program) to calculate square roots like $\sqrt{7}$; however, your calculator will deal only with rational numbers. In fact, it deals only with terminating decimals, a rational number subset. The π and $\sqrt{7}$ that your calculator reports are 10-digit decimal approximations of those real numbers.

This calculator and computer restriction to decimals does not affect our calculations, and we will even develop a program to calculate powers like e^π, knowing that by "calculate" we mean that we will find decimal approximations, just as the calculator values of e and π are approximations. For us, then, we only have the 10-digit approximations $e^\pi = 2.718281828^{3.141592654} = 23.14069263$.

Although we will use simpler examples, the program we will develop in this chapter will carry out that kind of calculation.

RATIONAL POWERS WE CAN ALREADY CALCULATE

Before we introduce a program to calculate B^E for any exponent E, you should understand that you already can evaluate many rational powers. In Chapter 3 we addressed the problem of calculating with integral exponents, exponents like 5, 17, and -3. In Chapter 4 when we developed the square root algorithm, you added to these the exponent $\frac{1}{2} = .5$ since $\sqrt{x} = x^{1/2} = x^{.5}$.

By combining the programs of those chapters, you can then also calculate expressions like $7^{5.5}$, $3.14^{17.5}$, and $0.004^{-3.5}$.

Here is how to change the form of those expressions so that you can use the earlier programs or the keys they represent:

$$7^{5.5} = 7^{5+.5} = 7^5 * 7^{.5} = 7^5 * \sqrt{7}$$

$$3.14^{17.5} = 3.14^{17+.5} = 3.14^{17} * 3.14^{.5} = 3.14^{17} * \sqrt{3.14}$$

$$0.004^{-3.5} = \left(\frac{1}{0.004}\right)^{3+.5} = 250^{3+.5} = 250^3 * 250^{.5} = 250^3 * \sqrt{250}$$

In each case you would then multiply the integral power that you obtained from the program of Chapter 3 by the square root that you obtained from the program of Chapter 4 to get your answer. In the case of $7^{5.5}$, for example, you would have

$$7^5 * \sqrt{7} = 16807 * 2.645751311 = 44467.14229$$

We can do still more than this. We can enter a number, say, 16, in our square root program, then take the result and enter it in the same program. This would give us the square root of the square root of 16: $\sqrt{\sqrt{16}} = \sqrt{4} = 2$.

This process would apply to any number. For example, for 5, we might have

$$\sqrt{\sqrt{5}} = \sqrt{2.236067977} = 1.495348781$$

Just what is it that we're calculating here? If we represent those square roots by $\frac{1}{2}$ powers, we have $\sqrt{\sqrt{x}} = \left(x^{1/2}\right)^{1/2} = x^{1/4} = x^{.25}$.

Aha! We can calculate not just $x^{.5}$ but also $x^{.25}$. In exactly the same way we have

$$\sqrt{\sqrt{\sqrt{x}}} = \left(\left(x^{1/2}\right)^{1/2}\right)^{1/2} = x^{1/8} = x^{.125}.$$

In fact, with a suitable number of square roots we can calculate any power whose exponent is itself a power of $\frac{1}{2}$. Thus we can calculate with the following exponents: .5, .25, .125, .0625, .03125, .015625, .0078125, and so on.

Recall at this point that the program we developed in the last chapter justified our use of the $\boxed{\sqrt{}}$ key on our calculator. We will use it from now on. If, for example, you want to calculate $347^{.125}$, you can simply enter (note that parentheses are omitted here and in similar situations.) $\boxed{\sqrt{}}$ $\boxed{\sqrt{}}$ $\boxed{\sqrt{}}$ 347. When you then press $\boxed{\text{ENTER}}$, your calculator should display 2.077500727. You can check that this is indeed the same as $347^{.125}$ by entering $347\boxed{\wedge}.125$.[3]

We will see that this ability to calculate such powers is the key to the rational power program that we will develop.

BINARY NUMBERS BETWEEN ZERO AND ONE

We talked about binary numbers for the whole numbers in Chapter 2; in this chapter we will make use of binary numbers between zero and one.[4]

Just as .1 represents the decimal number one-tenth, that same symbol, .1, represents one-half in binary. The key to understanding a general binary

---

[3] In doing this we are, of course, getting ahead of the game. It is exactly such calculations that we will introduce later in this chapter. At this point such a check merely assures us that we are on the right track.
[4] What should we call these numbers? Binimals? Bicycles? I know of no standard name for them.

like 1001.101 is to observe that the place values continue to the right as in the following display:

2^3	2^2	2^1	2^0		2^{-1}	2^{-2}	2^{-3}
1	0	0	1	.	1	0	1

The decimal values corresponding to these places are

$$8 \quad 0 \quad 0 \quad 1 \qquad \tfrac{1}{2} \quad 0 \quad \tfrac{1}{8}$$

or, converting to decimal fractions:

$$8 \quad 0 \quad 0 \quad 1 \qquad .5 \quad 0 \quad .125$$

Summing these values, we have $1001.101_{two} = 9.625_{ten}$.

Now if you focus attention on those values between zero and one, you should see that the decimal equivalents of those binaries are exactly those values that we found in talking about roots in the previous section. They are .5, .25, .125, .0625, .03125, and so on.

That suggests a way to think about calculating such numbers. Consider the following carefully chosen example: $3^{.625}$. We can rewrite $3^{.625}$ as $3^{.5+.125}$, and then (again recalling the rules for exponents) rewrite this as $3^{.5} * 3^{.125}$. Ah, but this is the same as $3^{1/2} * 3^{1/8}$, and you know from the last section that this is $\sqrt{3} * \sqrt{\sqrt{\sqrt{3}}}.^5$

This you can calculate. You can key

$$\boxed{\checkmark}\ 3\ *\ \boxed{\checkmark}\ \boxed{\checkmark}\ \boxed{\checkmark}\ 3\ \boxed{\text{ENTER}}$$

to find that $3^{.625} = 1.987013346$. (You can also check that result by entering $3\boxed{\wedge}.625$.)

There is something else you should notice about this particular calculation. If we write the exponent .625 in binary, we would have $3^{.101}$, and that exponent tells us where the square roots appear in our calculation:

$$3^{.101} = 3^{1/2+0/4+1/8} = 3^{1/2} * 3^{1/8} = \sqrt{3} * \sqrt{\sqrt{\sqrt{3}}}$$

That binary form also tells us how to proceed in the following way. Beginning with $.1_{two}$, we needed a single square root. Then we increase

---

^{5}In many of the descriptions that follow, the $\sqrt{\ }$ symbols are retained to clarify the issues involved, whereas in the program processing they would be calculated and the resulting decimal values carried forward.

the number with each place as we move to the right, including the factor in our product only when a 1 appears. Without the intermediate steps, then, here is how a more complicated binary power would work without those intermediate steps:

$$7^{.01011} = \sqrt{\sqrt{7}} * \sqrt{\sqrt{\sqrt{\sqrt{7}}}} * \sqrt{\sqrt{\sqrt{\sqrt{7}}}}$$

That exponent was, of course, in binary. We can change it to decimal form: $.01011_{two} = \frac{1}{4} + \frac{1}{16} + \frac{1}{32} = \frac{11}{32} = .34375_{ten}$.

Thus, if we carried out the calculation above, we would have calculated $7^{.34375}$.

It would be nice if all exponents were equivalent to products of powers of $\frac{1}{2}$. Unfortunately they are not, but, by using these exponents with care, you will see that we can approximate any power.

DEVELOPING AN ALGORITHM TO CALCULATE RATIONAL POWERS

At the outset you should notice that we can always separate the integer part of any power from the part less than one in the same way we did in the previous section. For that reason, we will focus on the calculation of rational exponents, E, with $0 < E < 1$. In other words, we want to calculate the values of expressions like: $37.5^{.709}$

Notice that the base can be any positive number;[6] only the exponent is restricted in this way.

Here is one way to think about calculating a power like the example we have shown:

1. Change the exponent to binary form.
2. Use square roots as shown in the last section to evaluate the expression.
3. Stop when sufficient accuracy is reached.

That is, in fact, what we will do but, once we develop a program, we will combine the first two of those algorithm steps. We will also need to see how to end the process if the exponent isn't exactly equivalent to a binary.

Let's see how we might accomplish step 1. We'll use that .709 exponent for our example. We could subtract .5 from .709 and use it as .1 in binary, and we could continue in this way, subtracting successive powers of .5.

---

[6]There is no problem with $B < 0$ for integer powers. We know, for example, that $(-2)^3 = -8$ and $(-2)^{-3} = -\frac{1}{8}$, but we could not assign a real number value to powers like $(-2)^{.37}$.

The process would look like this with the new digit in italics:

Decimal		Binary
.709		
$\underline{-.5}$ $\Rightarrow$		.*1*
.209		
.209		
$-.25$ won't go so $\Rightarrow$		.1*0*
.209		
$-.125$ $\Rightarrow$		.10*1*
.084		
.084		
$-.0625$ $\Rightarrow$		.101*1*
.0215		
.0215		
$-.03125$ won't go so $\Rightarrow$		.1011*0*
.0215		
$-.015625$ $\Rightarrow$		.10110*1*
.005875		
.005875		
$-.0078125$ won't go so $\Rightarrow$		.101101*0*

and so on.

There is an alternate process that leads to the same result. It is based on the fact that each time we double the remaining decimal number, the whole number resulting is the binary value we seek. Here is how this works for this same decimal .709:

$$.709 * 2 = 1.418 \Rightarrow .1$$
$$.418 * 2 = 0.836 \Rightarrow .10$$
$$.836 * 2 = 1.672 \Rightarrow .101$$
$$.672 * 2 = 1.344 \Rightarrow .1011$$
$$.344 * 2 = 0.688 \Rightarrow .10110$$
$$.688 * 2 = 1.376 \Rightarrow .101101$$
$$.376 * 2 = 0.752 \Rightarrow .1011010$$

and so on.

Notice that at each stage in this second process the whole-number part of the doubled product contributes to the binary representation and is removed from the decimal.

Now let's see how we could then carry out step 2 of that algorithm: to apply square roots to those binary positions in the exponent. Here again is the example with which we began this discussion: $37.5^{.709}$.

We have shown how to change that exponent (but not the base) to the binary value: $37.5^{.1011010...}$, so we should calculate

$$\sqrt{37.5} * \sqrt{\sqrt{\sqrt{37.5}}} * \sqrt{\sqrt{\sqrt{\sqrt{37.5}}}} * \sqrt{\sqrt{\sqrt{\sqrt{\sqrt{37.5}}}}} * \ldots$$

stopping only when we have our desired accuracy.

Instead of carrying out those processes separately, we will see that we can combine the process of converting the power to binary and taking those successive square roots. This leads to reasonably short programs.

TWO PROGRAMS

We will develop two programs to calculate B^E, with $0 < E < 1$. The first one follows the first process we used to convert the exponent to binary. I have numbered the lines in order to refer to them in the explanation that follows it:

```
    PROGRAM:RATPOW1
1   : Prompt B,E
2   : 1→P
3   : √B →S
4   : .5→X
5   : While E>.00000000005
6   :     If E≥X
7   :         Then
8                     E−X→E
9   :                 P*S→P
10  :     End (If)
11  :     X/2→X
12  :     √S →S
13  : End (While)
14  : Disp P
```

Because it is easy to get lost in following a complicated example, we'll consider a simple one, $7^{.625}$. We know that this is equivalent to $7^{.101}$ when the exponent is binary. Thus it should take only three passes through the `While` loop (one for each binary digit) to complete the calculation.

In line 1 we would enter $B = 7$ and $E = .625$. P is going to accumulate the power we seek; in line 2 we set that $P = 1$. The third line sets $S = \sqrt{7}$, the square root of the base, 7. In line 4 we set $X = .5$. That .5 is the first number that we attempt to subtract from the exponent in that first decimal-to-binary conversion algorithm. With these values established, we enter the `While` loop.

The `If` test asks if E is large enough to subtract $X = .5$ from it. Since $.625 > .5$, we perform this subtraction in line 8, leaving $E = .125$. But rather than wait until the entire conversion to binary is accomplished, we multiply P by $S = \sqrt{7}$ now. Thus we now have $P = 1 * \sqrt{7} = \sqrt{7}$.

Before leaving the `While` loop, we prepare ourselves for the next pass by replacing X by $X/2$ in line 11, which makes $X = .25$ and S by $\sqrt{S}$ in line 12, which makes $S = \sqrt{\sqrt{7}}$.

Although $E = .125$ is plenty to send us through the `While` loop again, it is not as large as $X = .25$, so the `If` test fails and lines 8 and 9 are not processed. But lines 11 and 12 are, and we end the loop with $X = .125$ and $S = \sqrt{\sqrt{\sqrt{7}}}$. E remains .125, so we enter the `While` loop a third time. In it $E = X = .125$, so the lines following `If` are processed. In line 8, E becomes 0, and in line 9, P becomes $\sqrt{7} * \sqrt{\sqrt{\sqrt{7}}}$. In line 11, S becomes $\sqrt{\sqrt{\sqrt{\sqrt{7}}}}$ and in line 12, X becomes 0.

Since $X = 0$, we are finished with the `While` loop and the value $P = \sqrt{7} * \sqrt{\sqrt{\sqrt{7}}}$ is reported.[7]

Now we'll examine a second program to calculate B^E, with $0 < E < 1$. This one follows the second procedure for changing the exponent from decimal to binary. I have again numbered the lines in order to refer to them in the following explanation:

```
     PROGRAM:RATPOW2
1    :Prompt B,E
2    : 1→P
3    : √B →S
4    : While E>0 and abs(S−1)>.00000000005
5    :     int(2E)→I
6    :     If I=1
7    :        P*S→P
```

[7] Notice that last value of S is never used. If after running this program you type S RETURN , you will obtain that fourth root of 7 that was retained.

```
8   :      2*E−I→E
9   :       √S →S
10  : End (While)
11  : Disp P
```

To calculate the same example we used in the other program, $7^{.625}$, we again enter $B = 7$ and $E = .625$ in line 1 and the values $P = 1$ and $S = \sqrt{7}$ are set in lines 2 and 3.

Notice that in this program, since E is being doubled in each step, it may never get small enough to meet the $E < .0000000001$ test of the While loop that we used in the other program. For that reason, an additional test is necessary. That other test involves S, which will be successively $\sqrt{B}, \sqrt{\sqrt{B}}, \sqrt{\sqrt{\sqrt{B}}}$, and so on. Because B is positive, these numbers approach 1.[8] In the example we have chosen, however, because E will be zero again after three passes through the loop, this additional test will not apply.

It is within the While loop that we use the second algorithm. In line 5 we double the exponent E to make $E = 1.25$ and take its integer value, 1, as I. This is exactly what we did to find the first binary place in that algorithm. But again as in the previous program, instead of accumulating the entire binary exponent, we process it now in the If instruction. Since $I = 1$, we multiply P by the current value of S, namely, $\sqrt{7}$. This makes $P = 1 * \sqrt{7}$.

Next, in line 8, since we have applied it to P, we subtract that $I = 1$ from the doubled value of E, making $E = .25$. And before leaving the While loop in line 9, we make $S = \sqrt{\sqrt{7}}$, again as in the other program.

The next time through the loop, doubling E produces 0.5, so the integer value of this is $I = 0$, and the If instruction is not processed. We end the loop with $E = 1$ and $S = \sqrt{\sqrt{\sqrt{7}}}$.

The last pass through the loop then makes $P = 1 * \sqrt{7} * \sqrt{\sqrt{\sqrt{7}}}$, just as we wish.

A GENERAL PROGRAM FOR RATIONAL POWERS

We will now combine our program in this chapter with the one in Chapter 3, modifying it and adding a few touches where appropriate, to construct a new program that calculates B^E for $B > 0$ and any rational E.

---

[8]When $B > 1$, as E approaches 0, B^E approaches 1 from above, but when $0 < B < 1$ as E approaches 0, B^E approaches 1 from below. That is the reason for using the absolute value of $S - 1$.

Instead of numbering individual lines in this program, I have numbered six sections:

```
     PROGRAM:RPOWALL
1    : Prompt B,E
     : 1→P
2    : 0→N
     : If E<0
     :     1→N
     : abs(E)→E
3    : int(E)→I
     : B→C
     : While I>0
     :     I−2*int(I/2)→R
     :     If R=1
     :          P*C→P
     :     C*C→C
     :     int(I/2)→I
     : End (While)
4    : E−int(E)→F
     : √B →S
     : .5→X
     : While F>.00000000005
     :     If F≥X
     :         Then
     :                  F−X→F
     :                  P*S→P
     :     End (If)
     :     X/2→X
     :     √S →S
     : End (While)
5    : If N=1
     :     1/P→P
6    : Disp P
```

Let's see how this program works.

In section 1 we enter our values for B and E and set $P = 1$. In section 2 we set $N = 0$ and change it to 1 only if E is negative. (We will not use N again until section 5.) Next we replace E by its absolute value, thus making E positive.

In section 3 we split off the integer value of E, calling it I, and send it through the program from Chapter 3 for integer powers. When this section is completed P represents the integer power of the base B.

Here it will be useful to remind ourselves how exponents work. Consider $7^{4.35}$: $7^{4.35} = 7^{4+.35} = 7^4 * 7^{.35}$. Notice that we multiply the integer power by the fractional power to obtain our answer. For this reason we simply continue with the value of P obtained from section 3 as we enter and process section 4, which is the same as our program RATPOW1, but with F replacing E.

When section 4 is completed, we then, have the value of B^E with positive E. Moreover, we know from section 2 whether the E with which we began the program is positive or negative. If $E < 0$, we know that we would have set $N = 1$. In that case we simply replace P with its reciprocal $1/P$. The answer is then displayed by section 6.

It is worth considering what happens in this program when special values of E are entered:

1. What if $E = 0$? In section 1, P is set equal to 1. In section 2, N remains 0. In the first line of section 3, I is set equal to 0 and therefore the While loop is skipped. In the first line of section 4, since both E and $\text{int}(E)$ are 0, their difference, F, is 0. This means that the While loop of this section is skipped as well. In section 5, since $N = 0$, P remains unchanged. Thus we display the answer, 1. Since $B^0 = 1$, we have the correct result.

2. What if $E = 1$? This time in section 3, $I = 1$, then $C = B$ and we enter the While loop. In it we calculate $R = 1$ (the remainder when 1 is divided by 2). Since $R = 1$, the If instruction gives us $P = 1 * C = C$, and the final instruction in the loop sets $I = 0$ (the quotient when 1 is divided by 2). This value of I means that the loop is not processed again. In section 4, $F = 1 - 1 = 0$ and then $S = \sqrt{B}$ and $X = .5$ (neither of which will be used). The second loop is not processed since $F = 0$ and the $P = C$ is displayed. Since we set $C = B$, this is the same as displaying B. Just what we want since $B^1 = B$.

3. What if $E = -1$? This would be just like processing $E = 1$, except that $N = 1$ in section 2. Thus in section 5, we would substitute $1/B$ for B. Okay again, since $B^{-1} = 1/B$.

Our RPOWALL program does indeed calculate B^E for $B > 0$ and any E. In doing so it carries out the function of the $\boxed{x^y}$ and $\boxed{x^{1/y}}$ keys on some scientific calculators and the $\boxed{\wedge}$ key on others.

6

LOGARITHMS

The invention of logarithms came on the world as a bolt from the blue.
No previous work had led up to it, foreshadowed it or heralded its arrival.
It stands isolated, breaking in upon human thought abruptly
without borrowing from the work of other intellects or
following known lines of mathematical thought.

—Lord Moulton

While today logarithms turn up in such applications as compound interest and radioactive decay, they were originally invented to simplify
calculation.

It is difficult now for all but elderly people like this author, who
lived before the advent of electronic calculators, to understand how much
time was absorbed in carrying out those calculations. An examination of
the notebooks of early scientists such as Newton and Kepler show how
detailed was their arithmetic. For example, it was not uncommon for them
to multiply numbers with 10 or even more digits, filling notebooks with
such calculations.

It is hard to imagine sitting down without a calculator to compute a
product like 73857 * 97836.

I just did so, and it took me more than 2 minutes to carry out the arithmetic. Once finished, I checked my result, 7,225,873,452, by calculator,[1]

[1]Notice that I kept this product within the 10-digit range of my calculator. If I multiplied
numbers with six or more digits by a calculator with 10-digit display, I would have had to

which took no longer than recording the digits, and was relieved to find that for once I had made no computation errors.

In fact, in precalculator times much science remained unexplored because of the time needed to carry out the lengthy mathematical operations required.

In order to address this calculation problem, Scottish mathematician John Napier and a few years later his English colleague Henry Briggs together developed logarithms much as we know them today. From the time of that invention in the early seventeenth century until the middle of the twentieth century, a period of over 350 years, this mathematical concept was used extensively and almost exclusively to simplify calculations.

The value of logarithms (logarithm is abbreviated log[2]) for computation derives from the fact that they "step down" arithmetical processes. Multiplication and division are calculated by addition and subtraction, powers, and roots by multiplication and division.

Here are the familiar properties:

$$\log(a * b) = \log a + \log b$$

$$\log(a/b) = \log a - \log b$$

$$\log(a^n) = n \log a$$

$$\log \sqrt[n]{a} = \frac{\log a}{n}$$

Of course, the key to using any of these calculation formulas—or, in fact, to apply logs in scientific settings—is determining the logarithms of the numbers.

An example will demonstrate calculation with logs. Let's carry out that multiplication I introduced earlier, $73857 * 97836$.

From that first defining equation, we know that $\log(73857 * 97836) = \log 73857 + \log 97836$.

We use our calculator's $\boxed{\log}$ key to obtain the values of the right side of that equation: $\log(73857 * 97836) = 4.868391663 + 4.990498688$.

---

apply some algebra to get the product. For example, I could calculate 1234567×7654321 by considering those numbers as binomials $(1234t + 567)(7654t + 321)$, with t representing thousand, calculating the four products and combining the results. Programming a computer to carry out such multidigit calculations and display them in groups of digits represents an interesting challenge. A program to do this is presented in Appendix L.

[2]The word *log* is used in two ways that are usually determined by the context. On many calculators $\boxed{\log}$ represents $\log_{10}$, that is, log with the particular base 10, but the abbreviation log is often used in a more general sense as meaning logarithm with any base. (Be aware, however, that on some calculators and computers log represents $\log_e$.)

We add those right side values to give us[3]

$$\log(73857 * 97836) = 9.858890351$$

We now have the log of the product. To get the product itself, we must find the antilog of 9.858890351; that is, we must "undo" the log function. More formally, we must apply the inverse function. That is accomplished by typing 2nd log 9.858890351 ENTER to produce 7225873445.

Aha! Our calculator has not given us the correct answer. Can you just see your seventh grade teacher marking a big check next to that result? "Partial credit?" you ask, and she simply frowns. You could argue that your answer is off by only 7 in over 7 billion. That is a relative error of less than .0000001%. And indeed that is as good as most 10-digit calculators can do by means of logs.[4]

Before calculators were available, there was no simple machine with a keystroke that would provide the logs of numbers. Instead the numbers were looked up in books of tables. Briggs, the coinventor of logarithms, produced a table of logs for the integers from 1 to 20,000 and from 90,000 to 100,000, accurate to 14 decimal places. (The missing 20,000 to 90,000 gap was later calculated by Adriaan Vlacq, a Dutch publisher.[5])

There were, however, slide rules, devices like the one shown in Figure 6.1, whose basic scales were marked off in logarithmic distances.

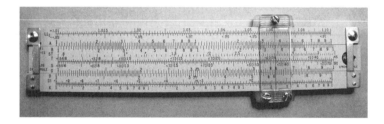

Figure 6.1   A slide rule.

[3]You would probably not have included that intermediate step, simply having typed log 73857 + log 97836 ENTER.

[4]It also suggests that calculator multiplication, which gave us an exact answer to that 10-digit product, is not done by logs.

[5]This and other interesting information about the early history of logarithms is found in Eli Maor, *e: The Story of a Number* (Princeton University Press, 1994). As Maor points out and as is so often the case, it may be that credit for this discovery has been assigned to the wrong person. A Swiss watchmaker, Joost Burgi, developed a system very similar to and arguably better than Napier's 26 years before Napier's publication of his results in 1614. Unfortunately for him, Burgi did not publish his ideas until 1620.

You can see how to develop a simple slide rule using log properties in Appendix P.

Their scales provide three- and sometimes even four-digit accuracy for computations. Until the late 1970s these wooden or metal devices were as widely used, especially among engineers, as scientific calculators are today. As soon as those calculators became available, however, slide rules (as well as most books of tables) became obsolete.

BASE 10 LOGARITHMS

log and 2nd log

Since some readers will have forgotten much information about logs, it will be useful to review what logs are before we develop programs to calculate them.

A key to understanding logs is the simple equivalence: *logarithm = exponent*.

In formal terms, you can see that equivalence in these two basic equations:

$$10^{\log N} = N \tag{I}$$

$$\log 10^N = N \tag{II}$$

Stated in prose, those equations tell us in two different ways that N is the power of 10 with exponent log N.

When you think of logarithms in those terms, you will find it simpler to derive many of their properties; for example

$$x * y = 10^{\log x} * 10^{\log y} = 10^{\log x + \log y}$$

The second equality sign (=) is based on the law of exponents $b^e * b^f = b^{e+f}$. Now if we take the logarithm of the first and last expressions, we have

$$\log(x * y) = \log(10^{\log x + \log y})$$

but the right side of that equation is, by equation (II), equal to log $x +$ log y. Thus we have derived the log equation: $\log(x * y) = \log x + \log y$.

Other log properties are derived similarly.

Return now to equation (I): $10^{\log N} = N$. From it we know immediately the logs of powers of 10. For example, suppose $N = 100$. Since $10^2 = 100 = 10^{\log 100}$, those first and last exponents are equal, and we have log $100 = 2$.

Similar substitutions give us a sequence of values:

$$\log 1000 = 3$$
$$\log 100 = 2$$
$$\log 10 = 1$$
$$\log 1 = 0$$
$$\log .1 = -1$$
$$\log .01 = -2$$
$$\log .001 = -3$$

For example, to show that $\log .01 = -2$, we think of the equation, $10^{\log .01} = .01$, which is the same as $10^{\log .01} = \frac{1}{100} = 10^{-2}$. Since the bases are alike (both 10), the exponents are equal and we have $\log .01 = -2$.

These integer values are called the *characteristics* of logs. If all we had to do was calculate with powers of 10, we could do so easily. What we need, however, are the values between those powers. For example, we need to find such values as $\log 3$, $\log 5280$, and $\log .57664$.

Before we turn to that problem, however, we need to recall some additional basics. Remember that in carrying out our example of log calculation, we came to the point at which $\log(73857 * 97836) = 9.858890351$.

We didn't want this log answer. Instead we wanted the actual value of $73857 * 97836$. Again we can use our defining relationship to determine this result: $73857 * 97836 = 10^{\log(73857*97836)} = 10^{9.858890351}$.

We could calculate $10^{9.858890351}$ by typing 10 $\boxed{\wedge}$ 9.858890351 $\boxed{\text{ENTER}}$ or on a scientific calculator using the $\boxed{x^y}$ key with $x = 10$ and $y = 9.858890351$ or using the program of Chapter 5. We need not do so, however, because the $\boxed{2^{\text{nd}}}$ $\boxed{\log}$ sequence gives you this value. In fact it is labeled "10^x" on most calculators.

What you should take from all this are the facts that calculating with logs involve:

1. Converting numbers to the corresponding powers of 10—that is, finding their logs via the $\boxed{\log}$ key
2. Using the rules for exponents to process the results
3. Returning to the answer by using the $\boxed{2^{\text{nd}}}$ $\boxed{\log}$ key sequence—that is, using the antilog or "10^x" key

Thus, a last way of seeing this calculation is $73857 * 97836 = 10^{4.868391663} * 10^{4.990498688} = 10^{9.858890351} = 7225873445$.

If you carried out those steps, using the $\boxed{\log}$ key to find the exponents in the second step, adding those exponents in the third, and applying the $\boxed{2^{\text{nd}}}$

log or "10^x" keys for the final step, you will have seen how logarithms carry out calculations.

NATURAL LOGARITHMS

When you think about it, the use of the base 10 is not necessary to work with logarithms. For calculations, however, 10 has advantages. In particular, you can tell the characteristic by inspection. For example, since 385 lies between 100 and 1000 or 10^2 and 10^3, its characteristic is 2. Beyond that, the decimal part of the log of 385, its so-called *mantissa*, will be the same as that of 38.5 or 385,000. A quick check with the log key will show these values[6]

$$\log 385 = 2.58546073$$

$$\log 38.5 = 1.58546073$$

$$\log 385,000 = 5.58546073$$

Whereas base 10 has these advantages for calculation, another base proves more useful in many problem settings, especially those related to exponential growth and decay and compound interest. That base is the irrational number $e = 2.718281828\ldots$[7] (While those repeated 1828s make it appear that a pattern has developed, it has not. If we continue the representation, the next two digits are 4 and 5. They do, however, make the number easier to remember.) Logarithms with this base are called *natural logs* and are found by use of the ln key.[8]

---

[6]A problem arises, however, with negative characteristics. You will find, for example, that your log key will give you $\log 0.0385 = -1.41453927$ instead of the -2.58546073 that you might expect, knowing that $\log 385 = 2.58546073$. This problem is addressed later in this chapter.

[7]The irrational number e has a formal definition, namely

$$e = \lim_{n \to \infty} \left(1 + \frac{1}{n}\right)^n$$

which to many people is confusing at best. One way of thinking about it is that $1,000,000e$ is the amount you would receive in return in one year for $1,000,000 invested at 100% rate of interest compounded continuously. I have used $1,000,000 * e$ because the bank would round $1 * e$ down to $2.71. (If you know where you can receive that interest rate not involving organized crime, please inform this author.)

[8]The numbers e and 10 are not the only useful log bases. Especially of value in computer science is $\log_2$, sometimes abbreviated lg. Once you have access to one log base, however, you can easily convert to another using a change of base formula, which will be discussed later and proved in Appendix N. An interesting applicaion that involves all three bases is in Appendix O.

To carry out the same calculation we did with $\log_{10}$ with ln, since $\ln 73857 = 11.20988607$ and $\ln 97836 = 11.49104789$, you would have $73857 \times 97836 = e^{11.20988607} \times e^{11.49104789} = e^{22.70093396} = 7225873476$. The calculation $e^{22.70093396} = 7225873476$ is found by using [2nd] [ln] 22.70093396. Notice that this approximation to the exact answer 7225873452 is different from the answer we achieved using $\log_{10}$ and again different from the exact answer.

DEVELOPING A PROGRAM TO CALCULATE LOGARITHMS

It turns out that, like trigonometric functions, on many calculators and computers logs are calculated by means of CORDIC-based programs, one of which is discussed in Chapter 9. In this section we will consider a slower program that is based on a simple relationship between numbers and their logarithms. We will use base 10 logs throughout this section but, once we have developed a program, we will see in a later section how to modify it for any other base such as e or 2.

Our program will be divided into two parts. In the first part we will calculate the characteristic, the whole-number part of the log; in the second part, the mantissa, the decimal part. To simplify this discussion, for now we will consider only logrithms of numbers greater than or equal to one. (We will address the problem of calculating logs of numbers between zero and one later.)

Suppose, for example, that we seek log 500. Here are program lines that will locate this log between two nonnegative integers, K and M:

```
: Prompt X
: 1→W
: 0→M
: While X > W
:     W*10→W
:     M+1→M
: End (While)
: W/10→U
: M−1→K
```

Let's see how that program segment works. We enter 500 for X. In the second and third steps the program sets $W = 1$ and its log, $M = 0$. Next we encounter a `While` loop that is governed by comparing X with W. Clearly our X, 500, is > 1, the current value of W, so we enter the loop.

Within the `While` loop, we multiply W by 10 and increase its log, M, by 1. This corresponds to log $10 = 1$. This loop is repeated, giving successively:

While	W	M
Before entry	1	0
After 1st pass	10	1
After 2nd pass	100	2
After 3rd pass	1000	3

Notice that the loop is processed until $W > 500$. Since it was still true that $X > M$ ($500 > 100$) when the loop was entered the final time, we end this part of the program with $W = 1000$ and $M = 3$. Although $W > X$ and its corresponding M is also larger than the characteristic we seek, as we will see, they will prove useful to us.

In the final two steps we calculate $U = 100$ and $K = 2$. Notice that this locates X between U and W (with $100 < 500 < 1000$) and log X between K and M (with $2 < \log X < 3$.)

More generally, this program segment takes an input value, $X \geq 1$, and ends with $U \leq X < W$ and $\log U \leq \log X < \log W$.

Before we move on to calculate mantissas, I remind you of two basic properties of logarithms that we will use:

1. We have

$$\log \sqrt{x} = \frac{\log x}{2}$$

    To show this, consider the following: $\log a^2 = \log(a * a) = \log a + \log a = 2 \log a$; thus $\log a^2 = 2 \log a$. Now substitute $\sqrt{x}$ for a and we have[9] $\log x = 2 \log \sqrt{x}$. Finally, dividing each side by 2, we have the desired relation:

$$\log \sqrt{x} = \frac{\log x}{2}$$

2. Suppose that we already know two numbers and their logs. For example, assume that we know $\log x = r$ and $\log y = s$, and we

---

[9]You may recall that $\sqrt{a^2} = |a|$. In working with logs we deal only with positive bases with $|a| = a$; thus we need not be concerned with absolute value.

want to calculate $\log \sqrt{xy}$. Using what we just showed and the rule for the log of a product, we have:

$$\log \sqrt{xy} = \frac{\log x + \log y}{2} = \frac{r + s}{2}$$

You should notice something about this relationship. It involves our old friends, on the left side the geometric mean, and on the right side, the arithmetic mean. Thus the log of the geometric mean of two numbers is the arithmetic mean of their logs.

Now remember how we started this discussion. We agreed to work from two values that we already knew, x and y. Let's see what happens if we start with the $x = 100$ and $y = 1000$ that were less than and greater than the 500 of our example in locating the characteristic. We have

$$\log \sqrt{100 \times 1000} = \frac{\log 100 + \log 1000}{2}$$

Because $\log 100 = 2$ and $\log 1000 = 3$, we have $\log \sqrt{100000} = (2 + 3)/2$, and, since $\sqrt{100000} = 316.227766$, we find that $\log 316.227766 = 2.5$.

That may at first seem like the silliest calculation you have ever seen. What use is there for the log of 316.227766, even if it does equal a nice decimal, 2.5? Indeed, that log alone does very little for us. The point is, however, that it gives us an additional value squeezing in on the value whose log we seek. We now know the number whose log is 2, the number whose log is 3, and, in addition, the number whose log is 2.5. We can use the same technique to find the numbers whose logs are 2.25 and 2.75.

Suppose, then, that we seek the number whose log is 2.75. We can find it in this way:

$$\log \sqrt{316.227766 * 1000} = \frac{\log 316.227766 + \log 1000}{2}$$

$$\log \sqrt{316227.766} = \frac{2.5 + 3}{2}$$

Calculating the square root on the left side and simplifying the right, we have $\log 562.3413252 = 2.75$.

We can, of course, continue to use this technique to find more and more logs. Proceeding to do so without some system will do us little good, though, so we must work out a system to calculate a particular logarithm.

Here is how we can continue the process of calculating $\log 500$. From the program segment that calculated the characteristic, we already know that $2 = \log 100 < \log 500 < \log 1000 = 3$, so $2 <$

log 500 < 3. We also found that 2.5 = log 316.227766 < log 500 < log 1000 = 3, so 2.5 < log 500 < 3 and 2.5 = log 316.227766 < log 500 < log 562.3413252 = 2.75, so 2.5 < log 500 < 2.75.

So far, then, we have closed in to the point that we know log 500 is between 2.5 and 2.75. Continuing with the corresponding values, we have

$$\log \sqrt{316.227766 * 562.3413252} = \frac{2.5 + 2.75}{2}$$

$$\log 421.6965034 = 2.625.$$

This squeezes in a bit further: 2.625 = log 421.6965034 < log 500 < log 562.3413252 = 2.75, and we have 2.625 < log 500 < 2.75.

Continuing in this same way, we would have

$$2.6875 < \log 500 < 2.75$$

$$2.6875 < \log 500 < 2.71875$$

$$2.6875 < \log 500 < 2.703125$$

$$2.6953125 < \log 500 < 2.703125$$

$$2.6953125 < \log 500 < 2.69921875$$

$$2.697265625 < \log 500 < 2.69921875$$

$$2.698242188 < \log 500 < 2.69921875$$

$$2.698730469 < \log 500 < 2.69921875$$

$$2.698730469 < \log 500 < 2.698974609$$

$$2.698852539 < \log 500 < 2.698974609$$

$$2.698913574 < \log 500 < 2.698974609$$

$$2.698944092 < \log 500 < 2.698974609$$

$$2.698959351 < \log 500 < 2.698974609$$

$$2.69896698 < \log 500 < 2.698974609$$

$$2.69896698 < \log 500 < 2.698970795$$

$$2.698968887 < \log 500 < 2.698970795$$

$$2.698969841 < \log 500 < 2.698970795$$

$$2.698969841 < \log 500 < 2.698970318$$

$$2.698969841 < \log 500 < 2.698970079$$

$$2.698969960 < 500 < 2.698970079$$
$$2.698969960 < 500 < 2.698970020$$
$$2.698969990 < 500 < 2.698970020$$
$$2.698969990 < 500 < 2.698970005$$
$$2.698969997 < 500 < 2.698970005$$
$$2.698970001 < 500 < 2.698970005$$
$$2.698970003 < 500 < 2.698970005$$
$$2.698970004 < 500 < 2.698970005$$
$$2.698970004 < 500 < 2.698970004$$

and indeed, as you can check by using a calculator's $\boxed{\log}$ key, log 500 = 2.698970004.

I have shown only the result of all of those steps, each involving calculation of a geometric mean and an arithmetic mean and determining which side of the inequality to change, in order to show how slowly this procedure converges.

You will now see that our program to calculate logs will carry out exactly the steps we have described and worked out by direct calculation.

This program consists of two parts. We have already seen the first part, calculation of the characteristic. The second part that we will develop now will calculate the mantissa, the decimal part of the log. The independence of these two parts should be evident when we recall that knowing log 500 = 2.69897004, we also know that log 50 = 1.69897004, log 5000 = 3.69897004 and so on.

What we need to do to develop this second part of our program is to show how to carry out that long series of calculations. We need only assume what we know so far: that we seek better and better approximations to log X, designated L in the program, and that we know (from the search for the characteristic in the first part of the program) that $K \leq L < M$. Of course, when we begin this process, K and M are integers with $K + 1 = M$. We also know that $U \leq X < W$.

Here, then, are program steps that will continue this process:

```
· · ·
: While K  ≠  M
:      √(U*W)→V
:      (K + M)/2→L
:      If X > V
:          Then
:                    V→U
:                    L→K
```

```
:           Else
:                   V→W
:                   L→M
:     End (If)
: End (While)
: Disp L
```

To see how the steps mirror the process we went through mechanically earlier, consider how it would have processed the first step of that example, finding log 500. Here is what we would know from the earlier program segment as we enter this program section:

1. $X = 500$, the number whose log we seek
2. Because $100 < X < 1000$, $U = 100$ and $W = 1000$
3. Because $2 < \log X < 3$, $K = 2$ and $M = 3$.

First, we should see that each pass through the While loop represents one step in the process of narrowing the interval within which log 500 is located.

The two program lines that narrow the interval are

```
:     √(U*W)→V
:     (K + M)/2→L
```

Once these geometric and arithmetic means are calculated, the decision is made whether to reduce the number and log interval from above or to raise it from below; then the appropriate action is taken.

```
:     If X > V
:         Then
:                 V→U
:                 L→K
:         Else
:                 V→W
:                 L→M
:     End (If)
```

This If-Then-Else series of steps may be illustrated by two familiar activities:

The first is the true/false test. The If line is asking a question, in this case, "Is $X > V$?" There are only two answers to a true/false question, as there are here. True takes us to the Then instructions, false takes us to the Else instructions.

The second is the children's game "Twenty Questions" when played with numbers. In that game you seek a number located between, say, 1 and 100. A good strategy is to ask a question that will divide the numbers in half, for example, you might ask, "Is the number greater than 50?"[10] If the answer is "Yes," you know now that your number is between 51 and 100; if the answer is "No," you know that the number is between 1 and 50. An algorithm for this single step could be written as follows:

```
If N > 50
  Then 51 < N < 100
  Else 1 < N < 50
End
```

In the case of our program the If question $X > V$? determines whether we close in from below or above. A true answer leads us to close in from below, and those are the actions following Then. If the answer is false, we close in from above by taking the actions following Else. This is represented in the diagram of Figure 6.2.

What we have done in this program segment is an example of what programmers call *binary search* or *interval bisection*. This common technique is used widely by programmers, and an example of its use in finding a root of an equation is given in Appendix M.

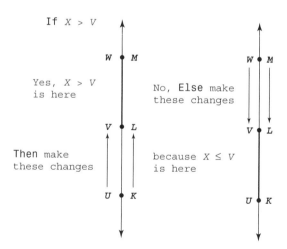

Figure 6.2 The If-Then-Else role in binary search.

[10]There are other possibilities to separate the numbers into two equal groups, for example, "Is the number odd?"

Notice that in both the Then and Else lines of our program two actions are taken. Both the number boundary and the corresponding log boundary are moved.

The While loop is repeated until the boundaries have squeezed so close that they approximately coincide; that is, when $K = M$, the process is complete. In fact, the While loop is repeating exactly those approximation steps shown on pages 93–94.

Here, then, is a program combining these two parts:

```
PROGRAM:LOGGRE1[11]
: Prompt X
: 1→W
: 0→M
: While X > W
:      W*10→W
:      M+1→M
: End (While)
: W/10→U
: M−1→K
: While K ≠ M
:      √(U*W)→V
:      (K+M)/2→L
:      If X > V
:           Then
:                     V→U
:                     L→K
:              Else
:                     V→W
:                     L→M
:      End (If)
: End (While)
: Disp L
```

There is much going on in this program. We know from the worked-out steps on pages 93–94 that if we enter $X = 500$, the second While loop alone must be processed 30 times before we achieve the answer, 2.698970004 for log 500. Thus it is not unexpected that this program is relatively slow. On my calculator it took almost 3 seconds to calculate this value. The $\boxed{\log}$ key, on the other hand, appears to produce an answer instantly.

Several important questions remain to be addressed.

---

[11]This name is an abbreviation for "log greater than or equal to 1."

LOGARITHMS OF SMALL NUMBERS

The program we developed in the last section considers only numbers $X \geq 1$. There is another restriction as well: we deal only with logarithms of positive numbers. The reason: logs are exponents and rational exponents on negative bases are useful only in advanced work.[12] We do want to know values like log .5 and log .00067, so we want to extend our program to calculate logs of numbers in the range $0 < X < 1$.

Notice that this problem involves calculation only of the characteristic for we know that, for example, the mantissa is the same for log .00067 as it is for log 6.7.

One way to address this problem would be to change our starting point in the first part of the program. We could, for example, replace the program lines

```
:  1→W
:  0→M
```

with

```
:  .00001→W
:  ⁻5→M
```

This program revision would take care of logs for numbers as small as .00001. Unfortunately, much of science involves numbers far smaller than that. It would, of course, be possible to set those starting numbers still smaller, but we seek a more general approach that will accommodate all positive numbers.

We will consider two solutions to this problem. In the first we treat negative characteristics differently from positive characteristics. Here is a program segment that separates calculation of the characteristic for $X \geq 1$ (what we did in the previous program) and $0 < X < 1$:

```
:  Prompt X
:  1→W
:  0→M
:  If X ≥ 1
:      Then
:              While X > W
:                      W*10→W
```

---

[12]Exceptions include a few powers like $(-27)^{1/3} = -3$. Others have complex roots like $(-8)^{1/3} = 1 \pm i\sqrt{3}$ as well as -2 (see Chapter 9), most of which require more advanced calculators or computers.

```
:                      M+1→M
:              End (While)
:              W/10→U
:              M−1→K
:       Else
:              While X < W
:                     W*.1→W
:                     M−1→M
:              End (While)
:              W→U
:              U*10→W
:              M→K
:              K+1→M
: End (If)
```

The change here is the addition of more If-Then-Else program lines, the Else lines taking care of values of X between 0 and 1.

Note that in that Else section the While loop descends:

```
:       While X < W
:          W/10→W
:          M − 1→M
:       End (While)
```

As the value of W is divided by 10, the corresponding characteristic, M, is reduced by 1.

Once W is less than X we must adjust the values of W, U, M and K so that $U \leq X < W$ and $K \leq \log X < M$. These lines are different from those at the end of the Then instructions because $W \leq X$ and $M \leq \log X$. The first instruction, W→U, gives us $U \leq X$, as we wish, and the second line, U*10→W, places W where we want it, immediately above X in numerical value:

```
: W→U
: U*10→W
: M→K
: K + 1→M
```

Then similar adjustments are made for M and K. As with W, M has overtaken $\log X$ and we have $M \leq \log X$. M→K corrects this to make $K \leq \log X$. Then the corrected value of M is found by adding 1 to this value.

The second approach is much simpler. In it we will determine a correction to be applied at the end of the program. Here are program lines that will do this. They will be inserted at the beginning of the program immediately after the instruction Prompt X:

```
: 0→C
: While X < 1
:      10*X→X
:      C+1→C
: End (While)
```

Notice first that this While loop would be bypassed if $X \geq 1$.

To see how this program segment works, consider how it would find the correction factor for log .005. Entering the While loop, $X = .005$, and $C = 0$, we would then have

	X	C
After pass 1	.05	1
After pass 2	.5	2
After pass 3	5	3

and then, because $X > 1$ we would leave the loop with $X = 5$ and $C = 3$.

These values would be used in the rest of the program. The (incorrect) value of X would produce a characteristic $K = 0$, and the While loop of that characteristic calculation section would be processed just once.

Then the mantissa would be calculated for 5 instead of .005, but we know that they would be the same. When we reach the final step of the full program, then, we have $L = \log 5$. We change that step from Disp L to Disp L−C to correct that characteristic.

Here is the complete program with these lines inserted (I leave the preparation of a complete program for the other solution to this problem to the reader):

```
PROGRAM:LOGX
: Prompt X
: 0→C
: While X < 1
:      10*X→X
:      C+1→C
: End (While)
: 1→W
: 0→M
: While X > W
:      W*10→W
:      M+1→M
```

```
: End (While)
: W/10→U
: M−1→K
: While K ≠ M
:        √(U*W)→V
:        (K+M)/2→L
:        If X > V
:            Then
:                      V→U
:                      L→K
:            Else
:                      V→W
:                      L→M
:     End (If)
: End (While)
: Disp L−C
```

In the example we considered log .005, and I claimed that the correction factor, $C = 3$, would give us this log instead of the log 5 that entered that last program line. If you were to run the program for log 5 and then for log .005 (or if you used the [log] key to determine these values), you would achieve the following results: log $5 = 0.6989700043$ and log .005 $= -2.301029996$.

Your immediate reaction may well be: "Wait a minute. Aren't those mantissas supposed to be the same? And isn't that characteristic of log .005 supposed to be -3?"

What has happened here is that your calculator has combined $-3 + 0.698970004$ to give -2.301029996. This is something that those of us who learned logs using log tables must get used to. In tables mantissas are always positive, so you end up with mixtures of negative characteristics and positive mantissas like that log $0.005 = -3 + 0.698970004$. For computational purposes, the calculator answer makes many problems much easier. For example, suppose that we want to calculate $x = \sqrt[4]{.005}$ by using logarithms:[13]

$$\log x = \log \sqrt[4]{.005} = \frac{\log .005}{4}$$

Thus we must divide log 0.005 by 4 and then find the antilog.[14] If, in that old world of table use, you did this problem, you would have been faced

---

[13]Of course, we could simply calculate the answer by using $0.005^{\wedge}(.25)$.

[14]To see this, let $x = \sqrt[4]{0.005}$. Then log $x = \log \sqrt[4]{0.005}$ and, by the properties of logarithms, log $x = (\log 0.005)/4$.

with
$$\frac{-3 + 0.698970004}{4}$$

Clearly, direct division would cause problems because that would give you a characteristic of $-.75$, which tells us nothing about the placement of the decimal point in the answer. The usual way of handling this situation was to make an adjustment to replace the characteristic by a number divisible by 4. Here, for example, you could change the problem to by subtracting 1 from the characteristic and adding it back to the mantissa,[15] to make this change:

$$\frac{-3 + 0.698970004}{4} = \frac{-4 + 1.698970004}{4}$$

Only now may the division be performed, giving $-1 + .424742501$. The antilog of $.424742501$ or $10^{.424742501} = 2.659147948$. Finally, moving the decimal point in this answer to represent the characteristic -1, you would have an answer, $.2659147948$.

Now consider the same problem with calculator-based logs

$$\log \sqrt[4]{.005} = \frac{\log .005}{4} = \frac{-2.301029996}{4} = -.5752574989$$

and antilog $(-.5752574989) = 10^{-.5752574989} = .2659147948$.

For applications to real-world problems, the calculator-calculated logs not only simplify but also clarify processing.

CALCULATING LOGARITHMS WITH OTHER BASES

The program we have developed calculates the logarithm with base 10. This is all that is needed to find a log to any base, because there is a formula for conversion from a base we know (in our case base 10) to any other base:

$$\log_B x = \frac{\log_A x}{\log_A B}$$

This formula is derived in Appendix N.

We will use the formula with $A = 10$,

$$\log_B x = \frac{\log_{10} x}{\log_{10} B}$$

---

[15]In school algebra, the procedure used is often still more complicated and involves addition and subtraction of ten before making further adjustments.

and since we have abbreviated $\log_{10}$ as log, we have

$$\log_B x = \frac{\log x}{\log B}$$

Through use of this formula we can now calculate logs with any base. For example, if we wish to find $\ln x$, that is $\log_e x$, we need only use

$$\ln x = \log_e x = \frac{\log x}{\log e}$$

We can determine $\log e$, that is, $\log 2.718281828$, by use of our program with $X = 2.718281828$. It would give us $\log e = 0.4342944819$. Instead of working this out for each calculation, we could merely substitute this value in the formula to give

$$\ln x = \frac{\log x}{0.4342944819}$$

or, since dividing by 0.4342944819 is equivalent to multiplying by its reciprocal,[16] 2.302585093, we have $\ln x = 2.302585093 * x$.

By an exactly similar procedure we could show that $\lg x$, that is, $\log_2 x$, may be found by the formula $\lg x = 3.321928095 * x$.[17]

Instead of following that route, however, we can develop a program to calculate $\log_B x$ for any base $B > 1$, by making very minor adjustments to our program for calculating $\log x$. Simply replace 10 by B throughout the program and change the input line to `Prompt B,X`.

To be complete, however, we need to take into account one final special case, the possibility that $0 < B < 1$.

We know that $B < 1$ implies $1/B > 1$, so let's see what happens if we use the change of base formula on $\log_{1/B} X$

$$\log_{1/B} X = \frac{\log_B X}{\log_B(1/B)}$$

and $\log_B(1/B) = Y$ is equivalent to $B^Y = 1/B$; thus $Y = -1$ and we have

$$\log_{1/B} X = \frac{\log_B X}{\log_B(1/B)} = \frac{\log_B X}{-1} = -\log_B X$$

---

[16]For anyone who has forgotten the meaning, the reciprocal of X is $1/X$ or X^{-1}. As used here, dividing by X is equivalent to multiplying by the reciprocal of X.
[17]You could add `Disp 2.302585093*X` or `Disp 3.321928095*X` to the end of the LOGX program to show the values of $\ln X$ or $\lg X$.

In other words, we can convert our base $B < 1$ to $1/B > 1$ as long as we change the sign of the resulting log. We can accomplish this with one last If instruction leading to one last change in the final program line. Here are the necessary program lines that follow Prompt B,X:

```
: 1→F
: If B < 1
:    Then
:        1/B→B
:        ⁻1→F
: End (If)
```

And we take into account this sign change in the final program line by changing it to

```
: Disp F(V − C)
```

This finally allows us to give a full program to calculate $\log_B X$ for any value of $X > 0$ to any base, $B > 0$:

```
PROGRAM:LOGBX
: Prompt B,X
: 1→F
: If B < 1
:    Then
:             1/B→B
:             ⁻1→F
: End (If)
: 0→C
: While X < 1
:     B*X→X
:     C+1→C
: End (While)
: 1→W
: 0→M
: While X > W
:     W*B→W
:     M+1→M
: End (While)
: W/B→U
: M−1→K
: While K ≠ M
:     √(U*W)→V
:     (K+M)/2→L
:     If X > V
```

```
:           Then
:                 V→U
:                 L→K
:           Else
:                 V→W
:                 L→M
:    End (If)
: End (While)
: Disp F(L − C)
```

When you run this program to find ln 500, you would respond to the cues:

$$\text{B?} \quad 2.718281828$$
$$\text{X?} \quad 500$$

Alternatively, you could respond to B? with $e(1)$. To find lg 500, you would respond with

$$\text{B?} \quad 2$$
$$\text{X?} \quad 500$$

FINDING THE ANTILOG

I conclude this chapter by answering a reasonable question: Do we need another program to carry out the inverse operation, finding the antilog?

We need only recall that the antilog$_B$ x is equivalent to B^x, to see that we need no additional program. In Chapter 5 we showed how to calculate rational powers. We need only use that program or the keys related to it with the given base to carry out the calculation.

For example, we have found that log $50 = 1.698970004$. If we want the antilog of 1.698970004, we should get 50. You can see if this is the case by entering $10^{\wedge}1.698970004$, by using the $\boxed{10^x}$ key with $X = 1.698970004$ or by using the program that justifies those keys in Chapter 5 with $B = 10$ and $E = 1.698970004$. The result for all three calculations should be very close to 50.

Similarly, if you want to find the antilog$_e$ X, the natural antilog, you would calculate e^x. If, for example, you knew that ln $Y = 5.521460918$, you could let $X = 5.521460918$, and enter $2.718281828 ^{\wedge} 5.521460918$ or use the e^x key with $X = 5.521460918$ or use the Chapter 5 program with $B = 2.718281828$ and $E = 5.521460918$. In every case you should find the result to be approximately 250.

7

ARCHIMEDES' CALCULATION OF π

Give me a place to stand and I will move the earth.

—Archimedes, quoted by Pappus of Alexandria (see Figure 7.1)

Figure 7.1 Archimedes' lever moves the earth (engraving from *Mechanics Magazine*, London, 1824.)

The electronics technology that supports modern computing is new, but the mathematics is not. In fact, some of that mathematics is almost as old

Inside Your Calculator: From Simple Programs to Significant Insights   By Gerald R. Rising
Copyright © 2007 John Wiley & Sons, Inc.

as civilization. In this chapter we will turn back the calendar 2200 years to see how Archimedes calculated π. It will turn out, as we will see in the next chapter, that his approach to this problem is mirrored by ours when we observe how cosine may be calculated.

Mathematicians consider Archimedes one of the half dozen or so finest members of their community in all history, belonging in the company of major figures like Newton, Kepler, Galileo, Euler, Einstein, and Gauss. It has been said that, if he were to return today, he would quickly adapt to modern terminology and become comfortable in any contemporary university mathematics department.

Unfortunately, like Rodney Dangerfield, mathematicians like Archimedes "get no respect" from the general public. The few who know anything about him have been led to consider him a figure of burlesque, rushing out naked from the Roman baths shouting, "Eureka! Eureka!"—*I have found it! I have found it!* Asked what he found, still fewer could respond. In that surely apocryphal story told by Vitruvius 300 years after Archimedes died, the mathematician had been challenged to determine whether a sacred wreath was made, as claimed, of pure gold. In the bath Archimedes noticed how his body displaced its volume of water, and he was able to apply this idea to solve the problem of the gold crown.

Important though the concept of displacement is to physics, it wasn't much of an accomplishment compared to what Sherman Stein calls Archimedes' "dazzling discoveries about the surface area and volume of a sphere, the center of gravity and the stability of floating objects." And he did much more. He contributed to military science, in particular designing catapults that exploited his principle of the lever. He developed the Archimedean screw, a pump for raising water. And, of course, he approximated the value of π to three-digit accuracy: 3.14. In fractional notation his values are $3\frac{10}{71} < \pi < 3\frac{1}{7}$.

That may not seem like much to us today until we recall that Archimedes had to work with the nearly useless pre-Hindu-Arabic numeration and had no easy means of calculating square roots. As I trace his mathematical development in what follows, I will largely bypass the additional problems these deficiencies created for him. You will see that what he accomplished, even forgetting the roadblocks, was extraordinary enough.

Recall that the number π is defined as the ratio of the circumference to the diameter of a circle. Thus $\pi = C/d$. This is, of course, equivalent to that formula learned in middle school, $C = \pi d$, by which the circumference of any circle is calculated from its diameter.

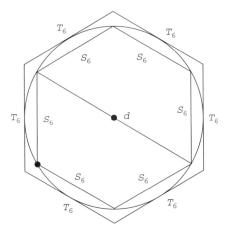

Figure 7.2 A circle with inscribed and circumscribed regular hexagons.

Archimedes begins his calculation of C/d for a fixed circle by squeezing the value of C between the perimeters of inscribed and circumscribed regular polygons, beginning with hexagons.

On the diagram in Figure 7.2 the circle has circumference C and diameter d. We denote the sides of the inscribed hexagon by S_6 and the sides of the circumscribed hexagon as T_6. (Throughout this chapter the subscripts designate the number of sides of the associated polygons.) The perimeter P_6 of the inscribed hexagon is then $6S_6 = P_6$, and the perimeter Q_6 of the circumscribed hexagon is $6T_6 = Q_6$. We then have the relationships $P_6 < C < Q_6, 6S_6 < \pi d < 6T_6$ and finally $6S_6/d < \pi < 6T_6/d$.

From this last inequality it should be clear that Archimedes had first to calculate single sides of those inscribed and circumscribed hexagons. Because all circles are similar, we can assume any size circle. We choose the unit circle, so named for having $r = 1$. Thus we have in that inequality $d = 2$. Archimedes then calculated the equivalent of $S_6 = 1$ and $T_6 = 2/\sqrt{3}$.[1]

Here is how this may be accomplished.

For S_6 the derivation is straightforward. The symmetries of the regular hexagon of Figure 7.3 allow us to divide it into six equilateral triangles, all of whose sides are equal to the radius of the circle. Thus, for the side of the inscribed polygon, we have $S_6 = r = 1$.

---

[1]School students often learn "simplifying" processes to rid fractions of irrational denominators. While this was quite reasonable when doing so avoided division by lengthy decimals, calculators have eliminated the need for this extra step. The technique of rationalizing the denominator (or occasionally the numerator) is, however, often useful in establishing equivalence of algebraic or trigonometric expressions.

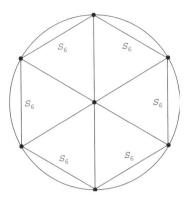

Figure 7.3

The side of the circumscribed hexagon requires a little more calculation. We have the polygon of Figure 7.4, divided into six equilateral triangles, but now the circle's unit radius is the length of a perpendicular bisector from the center to any of the sides of the regular polygon. (This segment is called the polygon's *apothem*.) To calculate T_6, we examine one of those triangles in Figure 7.5.

If we represent T_6 by $2x$ we can apply the Pythagorean theorem to one of the smaller triangles to give us $(2x)^2 = x^2 + 1^2$. Simplifying, we have $3x^2 = 1$, $x^2 = \frac{1}{3}$, $x = 1/\sqrt{3}$, and $T_6 = 2x = 2/\sqrt{3}$.

Archimedes was able to approximate this irrational number with the fraction $\frac{306}{265}$. We can calculate these values to compare them: $2/\sqrt{3} = 1.154700538$ and $\frac{306}{265} = 1.154716981$.[2]

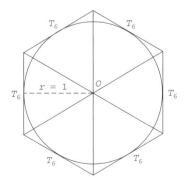

Figure 7.4

[2] Actually there is a still better estimate with a smaller denominator: 209/181. To see how to find such estimates, see Appendix K.

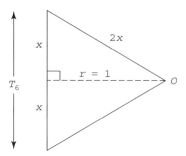

Figure 7.5

Notice that his value is accurate to 10-thousandths or five decimal places. How he derived this remarkably accurate estimation, he doesn't tell us and historians of mathematics have suggested several very complicated procedures as possibilities. We will not explore those avenues here.[3]

Recalling our inequality $6S_6/d < \pi < 6T_6/d$, we can now substitute our values for d, S_6, and T_6 to obtain

$$\frac{6(1)}{2} < \pi < \frac{6\left(\frac{2}{\sqrt{3}}\right)}{2} \quad \text{or} \quad 3 < \pi < \frac{6}{\sqrt{3}}.$$

Since $6 = 2 * 3 = 2 * \sqrt{3} * \sqrt{3}$, this may also be written $3 < \pi < 2\sqrt{3}$. This gives us a first approximation range for the value of π: $3 < \pi < 3.464101615$.

That one-digit accuracy doesn't tell us much, but it gave Archimedes, just as it gives us, a starting point. He now observes that doubling the number of sides of the inscribed and circumscribed polygons will give better estimates.

Consider first the case for the inscribed hexagon and dodecagon of Figure 7.6.

It seems clear that the sides of the dodecagon (12-gon) more closely approximate the circumference of the circle, but let us establish carefully that it does so, by considering what happens as one chord is replaced by two connecting the same endpoints.

In Figure 7.7 we have $S_6 < S_{12} + S_{12} < \text{arc}(AB) + \text{arc}(BC)$. Each of the inequalities is justified by the basic notion that the shortest distance between two points is the straight-line segment joining them.

---

[3]Several sources of further information about Archimedes, including speculation about his calculation of these square roots to such accuracy are included in the reading list at the end of this book.

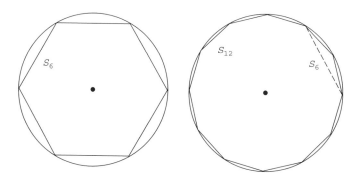

Figure 7.6

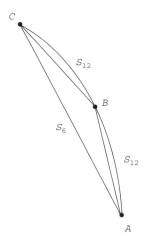

Figure 7.7

Simplifying this expression, we have $S_6 < 2S_{12} < \text{arc}(AC)$. We multiply through this inequality by 6 to obtain $6S_6 < 12S_{12} < 6 * \text{arc}(AC)$.

This inequality establishes that the perimeter of the hexagon, $6S_6$, is less than the perimeter of the dodecagon, $12S_{12}$, which in turn is less than the circumference of the circle, $6 * \text{arc}(AC)$, just what we want. We have shown that doubling the number of sides creates a polygon whose perimeter is closer to the circumference of the circle.

We turn next to the circumscribed hexagon and dodecagon of Figure 7.8 in order to establish that their perimeters "shrink" toward the circle circumference as the number of their sides is doubled.

Again it looks as though the dodecagon approximates the circle circumference more closely than the hexagon does, but let us justify that observation.

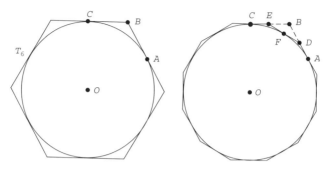

Figure 7.8

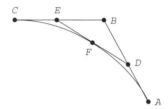

Figure 7.9

Instead of considering sides of the polygons directly, this time we focus on the length along the polygonal paths from A to C in Figures 7.8 and 7.9. This is one-sixth of the way around the circle. On the hexagon this is $AB + BC$, and on the dodecagon $AD + DE + EC$. Note that the symmetries of the regular polygons render $AB + BC = T_6$ and $AD + DF + FE + EC = 2T_{12}$. We must show that the second sum, which is equivalent to $AD + DE + EC$, is less than the first, $AB + BC$.

It turns out that we need consider only the points D and E. The shortest distance between them is clearly DE; thus $DE < DB + BE$. Add $AD + EC$ to each side of this inequality, and we have $AD + DE + EC < AD + DB + BE + EC$. Now combine $AD + DB = AB$ and $BE + EC = BC$ on the right side of the inequality to give us the desired $AD + DE + EC < AB + BC$. This is equivalent to $2T_{12} < T_6$.

Multiplying each side of this inequality by 6 gives us $12T_{12} < 6T_6$ and establishes that the perimeter of the dodecagon (the left side of the inequality) is less than the perimeter of the hexagon (the right side) and thus closer to the circle circumference.[4]

---

[4]Alert readers will note that I have not established that the perimeters of both circumscribed figures are greater than the circumference of the circle. Although it seems obvious (and is, of course, true), proving it here would take us too far from the basic argument. It is proved in Appendix O.

Although we have considered doubling the number of sides only in terms of hexagon to dodecagon, essentially the same arguments work whenever the number of sides is doubled. I encourage readers to see that this is the case by following our argument as it would apply going from inscribed and circumscribed squares to the corresponding regular octagons.

Archimedes did not bother with the justification we have gone through. He accepted it as true that doubling the number of sides improved his π estimates. Instead, he developed a procedure for calculating S_{2n} from S_n and T_{2n} from T_n. He had to do so by painstaking effort for each step, but the procedure he followed was always the same.

Beginning with the hexagon values that we have already calculated, which led him to simplify $6S_6/d < \pi < 6T_6/d$ to $3 < \pi < 2\sqrt{3}$, he was able to find the S and T values for polygons with 12, 24, 48, and 96 sides:

$$\frac{12S_{12}}{d} < \pi < \frac{12T_{12}}{d}$$

$$\frac{24S_{24}}{d} < \pi < \frac{24T_{24}}{d}$$

$$\frac{48S_{48}}{d} < \pi < \frac{48T_{48}}{d}$$

$$\frac{96S_{96}}{d} < \pi < \frac{96T_{96}}{d}$$

We can generalize these inequalities with the single statement: $nS_n/d < \pi < nT_n/d$, and we can also represent Archimedes' laborious steps by means of formulas, which turn out to be quite complex even in modern notation. The two formulas are[5]

$$S_{2n} = \sqrt{2 - \sqrt{4 - S_n^2}} \quad \text{and} \quad T_{2n} = \frac{(2\sqrt{T_n^2 + 4} - 4)}{T_n}.$$

To avoid interruption of this development, the derivation of these formulas is deferred to Appendix R.

---

[5]In order to avoid a square root within a square root, this expression may be rewritten

$$\sqrt{2 - \sqrt{4 - S_n^2}} = \frac{S_n}{\sqrt{1 + \frac{S_n}{2}} + \sqrt{1 - \frac{S_n}{2}}}$$

Proving this requires use of some algebra with which few mathematicians are familiar. It is included with the formula derivation of Appendix R.

While Archimedes had to work out in detail his individual values, we need not do this. We can instead set a calculator to run a program for us using our initial values and these formulas. Here is such a program:

```
PROGRAM:ARCHPI
: 6→N
: 1→S
: 2/√3 →T
: For (I,1,10)
:    Disp N,N*S/2,N*T/2
:    Pause
:    2N→N
:    √ (2−√ (4−S²))→S
:    (2*√ (T²+4)−4)/T→T
: End (For)
```

Some words of explanation are in order. N represents the current number of sides of the polygons. S and T represent the side lengths of the inscribed and circumscribed polygons beginning with the values we calculated for the hexagon ($N = 6$). Once these initial values have been set, the For loop first displays three things: the current number of sides and the result of the division perimeter/diameter (recall $d = 2$) for the inscribed and circumscribed polygons. These are the values between which π is squeezed. Then the number of sides is doubled (2N→N) and the new side lengths calculated by the conversion formulas before again displaying the new perimeter/diameter results.

Here are the results of running this program through its first five For loop passes together with correct significant digits underscored and an additional column indicating the number of correct digits at each stage:

I	N	P_N/d	Q_N/d	Accurate Digits
1	6	3	3.464101615	1
2	12	3.105828541	3.215390309	1
3	24	3.132628613	3.159659942	1
4	48	3.139350203	3.146086215	2
5	96	3.141031951	3.142714600	3

Archimedes went no farther than this, and his fractional approximations for square roots further restricted his results to $3\frac{10}{71} < \pi < 3\frac{1}{7}$, which may be represented to 10 digits (many of them carrying no meaning) in decimal form as $3.1408450704 < \pi < 3.1428571429$. Our program allows us to continue, however, until we obtain π to double that number of digits of accuracy. These represent the next five passes through the For loop.

I	N	P_N/d	Q_N/d	Accurate Digits
6	192	3.141452472	3.141873050	3
7	384	3.141557608	3.141662747	4
8	768	3.141583892	3.141610175	5
9	1536	3.141590463	3.141597036	5
10	3072	3.141592106	3.141593785	6

It is reasonable to modify our program to produce just that value for π accurate to seven digits. We need only move our `Disp` line to the end of the program, thus printing the value of π only once:

```
PROGRAM:ARCHPI2
: 6→N
: 1→S
: 2/√3 →T
: For (I,1,10)
:     2N→N
:       √(2-√(4-S²))→S
: End (For)
: Disp int(100000*N*S/2)/100000
```

Note that we have only calculated the values for the inscribed polygon, a further simplification of the program. That final program line illustrates a useful technique. It rounds the answer to five decimal digits by the following steps:

$$N * S/2 = 3.141592106$$
$$100000 * N * S/2 = 314159.2106$$
$$\text{int}(100000 * N * S/2) = 314159$$
$$\text{int}(100000 * N * S/2)/100000 = 3.14159$$

There are, however, limits to our program. Even if we increased the upper limit for I in the `For` loop, we would find ourselves no better off even with our powerful 10-digit calculators than Archimedes was with his severely limited resources. From here on our calculations lose accuracy because the small errors from rounding—even beyond the 10 to 14 our calculators carry—accumulate and lead to reporting inaccuracies. It is important to understand, however, that the program itself, if extended to a higher I limit, would continue to converge on the value of π when computers that process more digits are used.

Of course, we need only press the π key on our calculator to obtain a 10-digit value of π: 3.141592654. That value (which, as we have seen, includes additional digits) is simply stored in memory.

To recapitulate, we have seen how a brilliant mathematician over two millennia ago developed a systematic procedure for calculating π. We adopted his procedure, but we did so using the additional tools available to us largely because of our improved numeration and algebra. This allowed us to avoid Archimedes' extremely burdensome calculations.

But we must not forget that we shortcut some important geometry to make our story at least a bit easier to follow. If you wish to explore the story we have investigated further, you should visit Appendix Q to find how these complicated formulas for calculating the length of those new sides when their number is doubled may be derived:

$$S_{2n} = \sqrt{2 - \sqrt{4 - S_n^2}} \quad \text{and} \quad T_{2n} = \frac{2\sqrt{T_n^2 + 4} - 4}{T_n}.$$

You may also wish to see a different and quite remarkable approach to Archimedes' problem in Appendix S.

8

CALCULATING TRIGONOMETRIC FUNCTIONS

The analysis of angular sections involves geometric and arithmetic secrets which hitherto have been penetrated by no one.

—Francois Vieta

Recall now that seven-line program from Chapter 1 that calculated cosine:

```
PROGRAM:COSDEG
: Prompt D
: D*π/180→R
: R*R/4294967296→S
: For (I,1,16)
:    S(4−S)→S
: End (For)
: Disp 1−S/2
```

In this chapter we will see why that brief program works. But in order to do so, it is appropriate first to review some of the basic concepts of trigonometry. In doing so I will focus my discussion largely on the function cosine, since that is the function our program calculates. It is important to understand, however, that the other five circular functions are closely related to this one. Once we know cosine, we can calculate the others by means of the following formulas, determining the correct

Inside Your Calculator: From Simple Programs to Significant Insights   By Gerald R. Rising
Copyright © 2007 John Wiley & Sons, Inc.

sign from the size of angle A:

$$\sin A = \pm\sqrt{1 - \cos^2 A}, \qquad \csc A = \frac{\pm 1}{\sqrt{1 - \cos^2 A}}$$

$$\tan A = \pm\frac{\sqrt{1 - \cos^2 A}}{\cos A}, \qquad \cot A = \frac{\pm \cos A}{\sqrt{1 - \cos^2 A}}$$

$$\sec A = \frac{1}{\cos A}$$

You may or may not recall deriving those relationships from the basic definitions of the various functions. The point here is simply that our calculation of the single function cosine opens all of trigonometric calculation to us.

RIGHT-TRIANGLE TRIGONOMETRY

At some time in middle school most of us first met trigonometry. This initial contact with the ratios sine, cosine, and tangent was strictly in terms of that title word, because trigonometry clearly breaks down into *trig* = triangle and *metry* = measurement. We determined sides and angles of triangles through use of these ratios.

In fact, the subject at that level is even more restricted—to right triangles. On any right triangle, we learned, the cosine is the ratio between two sides:

$$\cos A = \frac{\text{adjacent leg}}{\text{hypotenuse}}$$

The sine and tangent have similar ratios:

$$\sin A = \frac{\text{opposite leg}}{\text{hypotenuse}}, \qquad \tan A = \frac{\text{opposite leg}}{\text{adjacent leg}}$$

In Figure 8.1 that means $\cos A = b/c$ and $\cos B = a/c$. The other functions follow similarly: $\sin A = a/c$, $\sin B = b/c$, $\tan A = a/b$ and $\tan B = b/a$.

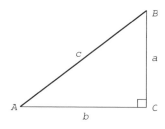

Figure 8.1

Those trigonometric ratios[1] allowed us to solve simple measurement problems in such fields as surveying and navigation.

THE CIRCULAR FUNCTIONS

Those definitions of cosine, sine, and tangent as ratios in a right triangle necessarily restrict use to angles between $0°$ and $90°$. Serious applications of these functions call for extension of these concepts. To do so we need to consider them as they are defined on a circle.

Strict usage requires us to call these extended definitions *circular functions* but, like the words *number* and *numeral*, which to some purists represent different concepts,[2] in general use, even mathematicians often refer to circular functions as *trigonometric functions*. I will do so here.

Clearly, we do not want the values of the trigonometric functions that we defined as right triangle ratios to be discarded as we move to new definitions. Instead we want to subsume those values in our new definitions.

Here is how that is done. We simply embed a right triangle, like the one on which we defined the trig ratios, in a unit circle—that is, a circle with radius one. We then center that unit circle on the coordinate plane with x and y axes as shown in Figure 8.2.

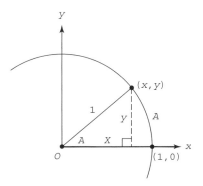

Figure 8.2   A point on the unit circle.

---

[1]The other trig functions, secant $x = 1/$cosine x, cosecant $x = 1/$sine x, and cotangent $x = 1/$tangent x, are defined similarly.

[2]When we talk of five marbles, we are talking about a number of marbles. A numeral, on the other hand, is a symbol for a number. Thus V and 5 are different numerals for the idea of five-ness. Back in the 1960s much was made of this by University of Illinois mathematics educator Max Beberman. In his UICSM algebra text Max had a student who was confused about number and numeral claiming that half of 8 is 3, "the right half." Clearly the student was talking about a numeral in that case.

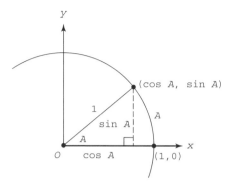

Figure 8.3   Circular function definitions.

On the diagram in Figure 8.2, our right triangle differs only in notation from the one we used for our trig functions. For continuity we will continue to use angle A now with its vertex at the origin O. By convention this angle A is considered positive when measured counterclockwise from the positive x axis. Because the radius of our unit circle is 1, the hypotenuse of our triangle is also 1. And finally, since we have chosen the point where the hypotenuse meets the circle to have coordinates (x, y), the horizontal side of the triangle is x and its vertical side is y.

It should be clear, then, that the trigonometric ratios on this triangle would be $\cos A = x/1 = x$, $\sin A = y/1 = y$, and $\tan A = y/x$. Also, since $\cos A = x$ and $\sin A = y$, we can modify our diagram by substituting these values wherever x and y occur. Figure 8.3 is the resulting diagram.

Many things are important about this new definition of these circular functions. Among them, A is no longer limited to acute angles; it can increase to $90°$ and beyond. As the angle increases, we continue the values of cosine and sine as the x and y coordinates of the point where the rotated ray meets the unit circle. Also, since the values of $\cos A$ and $\sin A$ are the coordinates of the point where the terminal side of the angle A meets the unit circle, we no longer need that right triangle. In fact, when A is a multiple of $90°$, there is no triangle, but we still have values. For example, when $A = 90°$, $x = 0$ and $y = 1$, thus $\cos A = 0/1 = 0$.

Further, the Pythagorean theorem establishes a basic trigonometric identity on the right triangle that will prove especially useful[3] $\sin^2 A + \cos^2 A = 1$.

The diagrams in Figure 8.4 show examples of these relationships for angles larger than $90°$. In order to insure that you follow this development,

---

[3]By convention $\sin^2 A$ represents $(\sin A)^2$.

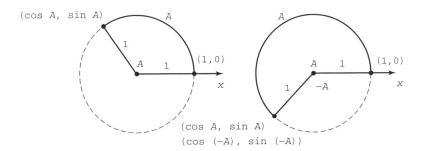

Figure 8.4   Circular functions beyond the first quadrant.

especially if you are meeting these ideas for the first time, you should sketch similar diagrams for such angles as $A = 0°$, $30°$, $45°$, $60°$, $120°$, $180°$, $225°$, and $330°$ and, by noting the x and y coordinates for those points, determine the values for cosine and sine for those angles.

DEGREES AND RADIANS

There is one additional point to be made about this diagram, and it is fundamental. Notice that A appears in two places. It labels both the angle at the origin and the arc between the point $(1,0)$ and the point $(\cos A, \sin A)$ on the circumference of the circle. Now clearly an arc and an angle are two quite different things. Why and how, then, do we make this association, and what is its purpose?

This correlation between the length of the arc and its angle (see Figure 8.5) is arbitrary but useful. In terms of degrees we need only think of one revolution as $360°$, whether it be the angle or the arc. A quarter way around, then, is $90°$, half way, $180°$, and so on.

Once that association is made, we follow it with a more fundamental association. We have made the arc for the complete circle $360°$, but on

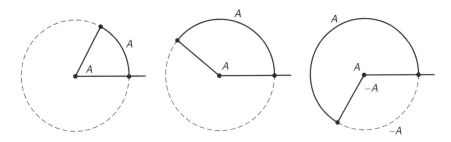

Figure 8.5   Angles and arcs on unit circles.

the unit circle the length of this same arc (the circumference of the circle) is 2π. On the basis of this association, we again arbitrarily make the connection $360° = 2\pi$, where the left and right sides of that equation represent different kinds of measure.

This is similar to our saying that 50 mm = 5 cm, but $360° = 2\pi$ has a more important role than do those measurement equivalents. The left side of that equation is clearly labeled in degrees, but the right side carries no corresponding label. The "measurement" assigned to that right side is often designated in radians, so our equation may be written $360° = 2\pi$ radians. There is a sense, however, in which this designation misses the central point of the association. The 2π arc length is a *length*, just as the radius is of length 1.[4] We haven't felt it necessary to call that radius 1 anything, and in particular few would even think of calling it 1 radian.

So we will stick with our unit circle relationship $360° = 2\pi$ except where clarity calls for use of the designation radian. We will see the power of this relationship in a moment, but first let us observe some of its consequences. We can, for example, divide each side of the equation by 4 to give us $90° = \pi/2$ or by 6 to arrive at $60° = \pi/3$ and so on. Another important relation is obtained when we divide that original equation by 2π. That gives us $\frac{180°}{\pi} = 1$. Carrying out that division on the left side tells us that about $57.3° = 1$. That this is quite reasonable is shown by Figure 8.6.

In Figure 8.6, notice that A is slightly less than the 60° angle of the dotted equilateral triangle, which further justifies our calculation of 1 radian $\approx 57.3°$.

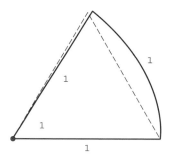

Figure 8.6

[4]If we wish to assign a unit of length to either the radius or the arc, then that same unit must be assigned to both. Thus a radius of 1 cm would have a 90° arc length of $\pi/2$ cm. On the other hand, we cannot do that with the equation $90° = 2\pi$: $(90°)$ cm $= 2\pi$ cm is meaningless because $90°$ cm carries no useful meaning.

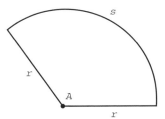

Figure 8.7   A sector.

Alternatively, we can also divide each side of our original equation $360° = 2\pi$, by 360 to give us $1° = \pi/180$. These two equations suggest the means of converting degrees to radians and vice versa. If our degree measure is D and radian measure is R, we have: $D * \pi/180° = R$ and $R * 180°/\pi = D$, or both relationships may be combined in the single proportion:

$$\frac{D}{R} = \frac{180°}{\pi}$$

Beware, however—the equivalence of $360°$ with 2π radians is only on the unit circle and that association does not extend to circles with other radii. On the sector of Figure 8.7, for example, it is straightforward to demonstrate through similarity that s, the length of the arc intercepted by the angle A on a circle with radius r, may be calculated by the formula $s = (A/360°)2\pi r = \pi r A/180°$, if A is given in degrees. More simply, if A is in radians, it is $s = rA$. Similarly, the area S of that sector in radians is $S = Ar^2/2$. These two formulas are important to engineering and other applications.

Now why all the fuss? This is, of course, the crux of the matter. We would not want to convert to radians unless it repays us in some way. And, indeed, it does provide a payoff. If we wish to perform calculations or explore relationships, we don't want to base our activities on differing units. Consider this in a rather simplistic context. Suppose that we wish to know the area of a rectangular path 42 cm wide and 9.6 m long. It would not be useful to recall the rectangle area formula, $A = lw$, and calculate $9.6 * 42$ to claim that our answer is about 400. Instead, we would wish to use common units, either centimeters or more likely in this case, meters. The area in meters would be $9.6 * .42 \approx 4.0$ square meters (m^2).

This seeming digression is important in relation to the cosine program that we seek to understand. As you will soon see, radians play an important role in that program.

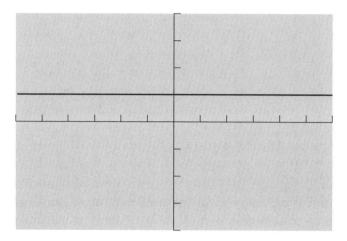

Figure 8.8   The graph of $Y = \cos X°$.

THE GRAPH OF COSINE

An amusing example of the problems associated with the use of degrees in equations involving circular functions results when inexperienced users of graphic calculators attempt to plot the graph of $\cos x$. They are surprised when they enter this function and press the $\boxed{\text{GRAPH}}$ key. What they obtain looks more like the graph of $Y = 1$ than the familiar wave of the cosine. (If you have a graphing calculator handy, you might wish to confirm this for yourself.)

The astonishing result shown in Figure 8.8 is due to the fact that they (or you, perhaps) have set up the calculator in `Degree` mode and have set the range of the x axis to a small number of units. As a result, the calculator plots values of cosine for angles very near $0°$. If you look back at the table of trigonometric values in Chapter 1, you will see that the values of cosine for small angles are all very near one.

As soon as you change the mode of the calculator to `Radian`, the graph will take its appropriate shape as, shown in Figure 8.9.

THE COSINE PROGRAM

One line of the cosine program with which this chapter began is necessary only if you enter a value of X in degrees. It is the line `X*π/180→X`.

That program line converts an angle X in degrees to angle X in radians. If you enter an angle in radians, that work is already done and the line would, in fact, create an error. Here, then, is the even simpler program for calculation of cosine R when R is entered in radians:

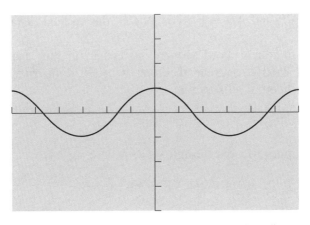

Figure 8.9   The graph of $Y = \cos X$ with X in radians.

```
PROGRAM:COSRAD
: Prompt R
: R*R/4294967296→R
: For (I,1,16)
:      S(4−S)→S
: End (For)
: Disp 1−S/2
```

Two other lines of the cosine program are central to the calculator calculation of cosine. They are S(4−S)→S and Disp 1−S/2. Once you see what those two steps are doing, you should understand the program.

Consider the second of those program lines first.

On unit circle O with diameter AB as shown on Figure 8.10, angle X or angle AOC has its equivalent arc X or arc AC. The coordinates of point C are then $(\cos X, \sin X)$. CD is drawn perpendicular to the diameter AB to form triangle ACD, and radius OC is drawn. On the diagram, $\cos X$ is the horizontal coordinate of point C. It is important to notice that $\cos X$ is negative, since D is to the left of the origin O. For that reason,

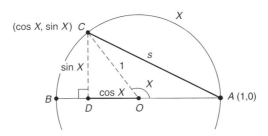

Figure 8.10

$-OD = \cos X,^5$ and $AD = 1 - \cos X$. CD, the vertical distance to point C, is $\sin X$.

We want to relate chord $s = AC$ to $\cos X$ on this diagram. To do so, we apply the Pythagorean theorem twice. In right triangle ODC the Pythagorean theorem tells us that

$$\cos^2 X + \sin^2 X = 1 \tag{I}$$

and it also applies to right triangle ADC where we have:

$$(1 - \cos X)^2 + \sin^2 X = s^2 \tag{II}$$

We can square $1 - \cos X$ using the identity $(a - b)^2 = a^2 - 2ab + b^2$: $(1 - \cos X)^2 = 1 - 2 * \cos X + \cos^2 X$. Substituting this in equation (II) gives us

$$1 - 2 * \cos X + \cos^2 X + \sin^2 X = s^2 \tag{III}$$

But from equation (I) we have $\cos^2 X + \sin^2 X = 1$, so we can replace these terms with 1 in equation (III) to obtain $1 - 2 * \cos X + 1 = s^2$, which may be solved for $\cos X$ by the following steps:

$$2 - 2 * \cos X = s^2$$

$$2 - s^2 = 2 * \cos X$$

$$1 - s^2/2 = \cos X$$

Here we make a key substitution. We set $s^2 = S$. When we do that, we have $1 - S/2 = \cos X$.

Since we began the program by entering X, we display its cosine with that last line instruction: `Disp 1-S/2`.

Be sure you see what this derivation tells us. It says that, if we can find the square of the length, s, of that chord AC, that is, s^2 or S, we can display the value of $\cos X$ simply as $1 - S/2$.

We are, of course, far from home free. We must now find a way of determining S, knowing X, the angle or arc whose cosine we seek. Unfortunately there is no direct way of doing this without applying the very trigonometric functions we seek to determine. We will instead find a way to approximate this relationship.

---

[5]This is an important point. OD represents a segment and a segment length is always a positive number. (You wouldn't say that the side of a triangle is -3, for example.) But here the cosine is measured to the left of the origin on the x axis and is therefore negative. That is why we must say $\cos x = -OD$.

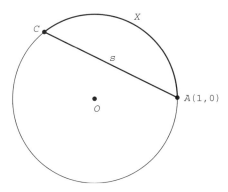

Figure 8.11

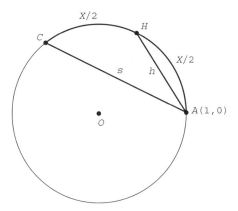

Figure 8.12

Figure 8.11 represents the same unit circle as the previous diagrams, but here we focus on $s = AC$ and arc X, eliminating the other lines and labels of that earlier diagram.

Now we bisect arc X, labeling its midpoint H and each half $X/2$. We also draw chord AH of length h, to produce Figure 8.12.

We next add two segments to the diagram, diameter HJ and chord AJ to obtain Figure 8.13.

Now more geometry theorems come into play. The first is a result of the symmetry of the circle. Because that diameter HJ is drawn from the midpoint of arc AC, it also bisects chord AC and is perpendicular to that chord. Thus we have M the midpoint of AC, with CM and AM both equal to $s/2$. AM is also perpendicular to HJ.

Another theorem tells us that any angle inscribed in a semicircle is a right angle. In Figure 8.14, for example, the angles at P, Q, and R are all right angles.

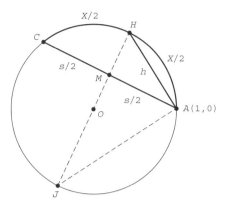

Figure 8.13

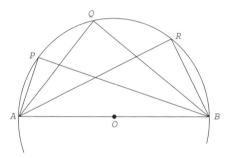

Figure 8.14

In our case this means that angle *JAH* is a right angle and in right triangle *JAH* we can apply the Pythagorean theorem to give us $HJ^2 = HA^2 + AJ^2$.

The length of diameter *HJ* is 2 since we are working in a unit circle. Substituting for *HJ* and HA then, we have $2^2 = h^2 + AJ^2$. Solving this for *AJ*, we have $AJ = \sqrt{4 - h^2}$. This gives us Figure 8.15.

We focus our attention on triangle *HAJ* which is extracted to give Figure 8.16.

Now we need only recall the area formula for any triangle: area $= \frac{1}{2} *$ base $*$ height. On this triangle we can use *HJ* as base and *AM* as height to give us

$$\text{Area} = \frac{1}{2} * 2 * \frac{s}{2} = \frac{s}{2}$$

We can also use *AJ* as base and *AH* as height to give us

$$\text{Area} = \frac{1}{2} * h * \sqrt{4 - h^2}$$

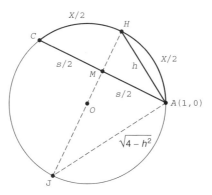

Figure 8.15

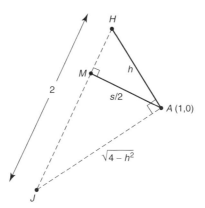

Figure 8.16

Now, clearly, since these two area results are for the same triangle, they are equal. That means that $\frac{s}{2} = \frac{1}{2}h\sqrt{4 - h^2}$ or $s = h\sqrt{4 - h^2}$.

That was a rather stiff argument, but we now have an important relationship that will be a key to our calculation. In order to avoid dealing with that square root, we can square both sides of the equation to give us $s^2 = h^2(4 - h^2)$.

We began this development by setting h as the chord of half the arc of chord s. This formula, however, tells us how to go from the smaller chord to the larger, so it is a formula for the increase in chord length as arc lengths are doubled. When we double the length of an arc, we can find the length of the chord of the new arc s by substituting the length of the shorter arc h in this formula. We will make use of this relationship shortly.

But first we must return to that original diagram on which we have the means to obtain our $\cos X$ if we know s or its square, S. But we don't know the chord s; instead all we know is X, the length of the arc AC.

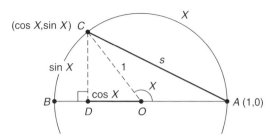

Figure 8.17 $\cos X = 1 - s^2/2$.

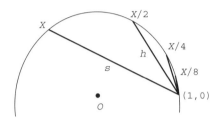

Figure 8.18

We have just gone to a great deal of trouble to show how to obtain the chord of an arc from the chord of an arc half as long (see Figure 8.17). We apply this procedure many times as on Figure 8.18. On this diagram the arc lengths are placed at the ends of the arcs.

We know now how to get from the chord for the arc of length $X/2$ (h on the diagram) to the chord of length X (s on the diagram). But we could use this same procedure to go from $X/4$ to $X/2$, from $X/8$ to $X/4$, and so on.

To see why we want to do this, note how the lengths of the chords are approaching closer and closer the lengths of their associated arcs. On the diagram it is already difficult to see the difference when we have halved the arc only three times. Although the chord will never equal its arc, we will make it as small as possible and then consider them close enough to be considered equal.

Just how small should we get? The answer to this question is determined by the number of digits of accuracy your calculator or computer manages. In the case of calculators displaying 10-digit accuracy, we can get very small, indeed. In fact, we can take half of the arc 16 times, so that we have an arc of length $X/2^{16} = X/65536$ before we make this exchange.

We can shorten our work by dealing with the squares of the chord lengths at each step. Recall in this regard that we found our relationship to be simpler when we used squares in the equation $s^2 = h^2(4 - h^2)$.

We start, then, with the square of that tiny chord $X/65536$. That is $X^2/4294967296$. And that is the source of that strange program line: $X*X/4294967296 \rightarrow S$.

Notice that we have stored this value in S. This is not, of course, the S that, we will use in our final program step. Instead, it is the square of this very small chord. We will use S for h^2 as well as s^2.

This means that we can take that formula, $s^2 = h^2(4 - h^2)$, convert it to an algorithm step

$$h^2(4 - h^2) \rightarrow s^2 \qquad (I)$$

and let S play the roles of both h^2 and s^2. This leads us finally to that program line: $S(4-S) \rightarrow S$.

What we must do now is apply this process for finding the new S as we double the chord length 16 times, finally

$$\frac{X}{2^{16}} \rightarrow \frac{X}{2^{15}} \rightarrow \frac{X}{2^{14}} \rightarrow \cdots \rightarrow \frac{X}{2^2} \rightarrow \frac{X}{2} \rightarrow X$$

or as we calculate at each step

$$S \rightarrow \text{new } S \rightarrow \text{new } S \rightarrow \cdots \rightarrow \text{new } S \rightarrow \text{new } S \rightarrow \text{new } S$$

ending up with the S corresponding to the square of the full arc length X. To do this, we use the For loop:

```
For (I,1,16)
   S(4-S)→S
End (For)
```

RECAPITULATION

To see how this all comes together, we will go through the program COSDEG step by step.

: Prompt X   Enter angle $X°$.

: X*π/180→X   This line converts $X°$ to radian measure.

: X*X/4294967296→S   This line accomplishes four things. It:
    (1) divides the arc X by 2^{16} (or 65536),
    (2) assigns that arc length to the corresponding chord length,
    (3) squares the result $(X/65536)^2 = X * X/4294967296$, and
    (4) assigns this squared value to S.

It is also in this (third and last) line that the jump from that short arc to its associated short chord is accomplished.

```
: For ( I, 1, 16)
:    S(4−S)→S
: End (For)
```

These three lines apply the formula $h^2(4 - h^2) \to s^2$, 16 times with S alternately playing the role of both h^2 and s^2. Recall that in the step before this loop, S was already the square of that chord of length $X/65536$. When the loop is ended, the S is the square of the chord for the arc X:

```
: Disp 1−S/2
```

Finally we apply that formula $1 - s^2 = \cos x$ to display the value we seek.

TRACING THE COSINE PROGRAM

I have made much in Chapter 1 and in this chapter of the brevity of the program for calculating the cosine of an angle expressed in degrees. It is useful, however, to see how much calculation goes on as those few lines are processed.

Suppose, for example, that we seek $\cos 128°$. When we enter the seven program lines in our calculator or computer and run the program, it simply carries out its responsibilities and prints the result: $-.6156614754$. What we will examine in this section are the intermediate steps in that calculation. Here is the cosine program modified to display intermediate values.

```
PROGRAM:COSXDISP
: Prompt X
: X*π/180→X
: Disp X
: Pause
: X*X/4294967296→S
: Disp S
: Pause
: For (I,1,16)
:    S(4−S)→S
:    Disp S
:    Pause
: End
: Disp 1−S/2]
```

The `Prompt` `X` command displays on the screen X=? and we would type 127 ENTER. This sets $X = 128$.

The next program line, `X∗π/180→X`, takes the current value of X (in this case 128), multiplies it by π, divides the result by 180, and stores the result back in X (in place of the 128). As a result, now $X = 2.234021443$.

In the next step, `X∗X/4294967296→S`, this value of X is multiplied by itself, the result divided by 4294967296, and the quotient stored in S. For our example, the result of this calculation is. 000000001162023238, obviously a very small number. (If you adapt the program to print intermediate values, this would be displayed in scientific notation, as 1.162023238E-9.) In any case we now have (suppressing any hidden digits) $S = .000000001162023238$.

This value of S is taken into that `For` loop and run through the step `S(4−S)→S` 16 times. Here are the step-by-step results displaying the decimal values. The values show how the square of the chord length increases each time the arc length is doubled:

Step	Decimal Value
1	.000000004648092952
2	.00000001859237179
3	.0000000743694868
4	.0000002974779417
5	.000001189911678
6	.000004759645297
7	.00001903855853
8	.00007615387166
9	.0003046096872
10	.001218346
11	.0048718995
12	.0194638625
13	.0774766081
14	.3039038077
15	1.123257707
16	3.231322951

That is quite a bit of work for three lines of program. Once again you should see the importance of calculating speed and the power of `For` loops.

Finally, in that last line, `Disp` `1−S/2`, the arithmetic gives us the displayed value for $\cos 128°$:

−.6156614754

So we see once again that much goes on inside your calculator when you run a program.[6] I doubt that any reader would enjoy carrying out this series of computations on a nonprogrammable calculator, to say nothing of trying to do this arithmetic by hand, even though no step involves arithmetic beyond subtraction, multiplication, and division. Here, for example, is the arithmetic you would have to perform just to convert from degrees to radians: $128 * 3.1415926535898/180$.

AN ASIDE ABOUT THE COSINE CALCULATION

I consider this cosine program a favorite for a number of reasons:

1. It is brief and for that reason alone quite elegant.
2. It calculates a significant result that earlier but within my lifetime could not be done.[7]
3. The rationale for the program derives from elementary geometry and few of us ever see geometry used in two such perfect settings.
4. The program makes use of a quite remarkable but reasonable trick: the switch from arc to chord for very short arcs.

The best way for anyone to gain full understanding of a lengthy development like that of this chapter is to repeat the derivation in a different setting. The program was developed here for a second quadrant angle. You might try to go through the derivation for an acute angle.

Finally, I must remind you of a point made back in Chapter 1. Programming speed is very important to programmers and this program, fast as it is, is still not fast enough for professionals. They calculate the trig functions by a quite different means that is the subject of the next chapter.

A PROGRAM TO CALCULATE ARCCOSINE

The program that we have developed will produce y from a given x in the equation $y = \cos x$. We conclude this chapter by addressing the problem of producing x from such an equation, given y. This is the inverse of

---

[6]We will see that the CORDIC program that carries out these calculations is equally complex. We will meet it in the next chapter.

[7]I am prepared to hedge here only slightly. It is within the realm of possibility that individual results could be obtained through days of hand calculating drudgery. In fact, exactly that kind of drudgery took up most of the waking hours of scientists like Johannes Kepler. That they finally accomplished results as accurate as those we are able to obtain in seconds I believe remains a remarkable achievement.

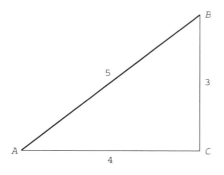

Figure 8.19

the function cosine, just as the square root function is the inverse of the function squaring.

This inverse function is designated $\cos^{-1}$ or arccos.[8] It is widely used in mathematics, but a simple example will show an elementary application.

Suppose that you want to know the smallest angle of the 3−4−5 right triangle of Figure 8.19.

We know that $\cos A = 4/5$, which is equivalent here to $A = \cos^{-1}(4/5)$.

A program that will calculate arccosine will reverse the process of our program to calculate cosine. Here it is:

```
PROGRAM:ACOSDEG
: Prompt X
: 2(1−X)→S
: For (I,1,16)
:     S/(2+√(4−S))→S
: End (For)
: 4294967296*S→T
: √(T)→R
: 180*R/π→D
: Disp D
```

To compare the two programs, I will place the corresponding lines side by side and consider them separately. Note that in these pairings the arccosine segments are given in reverse order.

```
PROGRAM:COSDEG          PROGRAM:ACOSDEG
: D*π/180→R              : 180*R/π→D
```

---

[8]Possible confusion is caused by the use of −1 here (and elsewhere) to indicate the inverse function. In other settings it represents the reciprocal, as in $x^{-1} = 1/x$. The expression $(\cos x)^{-1} = 1/\cos x$ should not be confused with $\cos^{-1} x$, the inverse cosine.

The first pair represent changing degrees to radians and radians back to degrees, since $(D * \pi/180) = R$ is equivalent to $(180 * R)/\pi = D$.

```
PROGRAM:COSDEG              PROGRAM:ACOSDEG
: R²/4294967296→S          : 4294967296*S→T
                           : √ (T)→R
```

The same is true of this pair, except that T plays an intermediate role. Indeed, the two statements of the arccosine program could have been combined into one:[9]

```
: 65536*√ (S)→ R
```

Note that $\sqrt{4294967296} = 65536$. Again the lines represent the reverse conversion. In the cosine program R is squared and the result divided by that large number. In the arccosine program R is multiplied by that large number and the square root taken of the result:

```
PROGRAM:COSDEG              PROGRAM:ACOSDEG
: For (I,1,16)             :For (I,1,16)
:   S(4-S)→ S             :    S/(2+√ (4-S))→ S
: End (For)                : End (For)
```

The most complex reversal is in these For loops. The arccosine program must reverse the process of S(4-S)→S. To do that, we must solve the equation $Y = S(4 - S)$ for S. This is equivalent to solving the quadratic $S^2 - 4S + Y = 0$.

The quadratic formula with $a = 1$, $b = -4$ and $c = Y$ provides the following solutions: $S = 2 \pm \sqrt{4 - Y}$. In order to keep the answer within the appropriate range, we must choose the negative sign. When we select that and replace Y by S, we have the instruction $2 - \sqrt{4 - S} \to S$, but this line is still problematic. When S is small, the value of the radical is nearly 4, making the square root close to 2. When this, in turn, is subtracted from 2, the result can have only a few significant digits, leading to inaccurate computation results.

To address this problem, we again, as in Appendix R, rationalize the numerator as follows:

$$2 - \sqrt{4 - S} * \frac{2 + \sqrt{4 - S}}{2 + \sqrt{4 - S}} = \frac{4 - (4 - S)}{2 + \sqrt{4 - S}} = \frac{S}{2 + \sqrt{4 - S}}$$

---

[9]The 4294967296 is retained simply to show the same number in the program.

Because now we are adding the value of the square root, the problem of getting too near zero no longer applies. Thus we have the resulting instruction:

S/(2+√(4−S)) →S

```
PROGRAM:COSDEG          PROGRAM:ACOSDEG
: Disp 1−S/2            : 2(1−X)→S
```

Set $1 - S/2 = X$ and solve for S, and you find this final equivalence.

While this program calculates $\cos^{-1} x$ in degrees, a simple change causes the program to display the answer in radians. You need only replace the final two lines of the program with : Disp R.

The entire program is

```
PROGRAM:ACOSRAD
: Prompt X
: 2(1−X)→S
: For (I,1,16)
:     S/(2+√(4−S))→S
: End (For)
: 4294967296*S→T
: √(T)→R
: Disp R
```

9

CORDIC CALCULATION OF COSINE

> Many things have more than direction; The magnitude is also a question.
> With acceleration or force, And many more things, of course,
> It's vectors that make the connection.
>
> —Jan Gullberg

A note at the outset: You have now explored how your calculator *can* be programmed to perform various key functions. This chapter readdresses one of those functions, cosine, from a different, more complex and technical perspective. In doing this it comes close to the way many calculator functions *are* programmed by computer scientists. This chapter should serve, then, as an introduction to the real world of calculator engineering. In the process it will provide further insights into calculator programming and important precalculus mathematics.

You have seen how simple programs can carry out the functions of calculator keys. But a reasonable question is: Do the calculator keys really work the way I have modeled them in earlier chapters?

Although calculator companies are often unwilling to divulge trade secrets, you can find in technical manuals some partial answers to this question, and that answer is, in the case of several functions you have met here, quite simply "No". What you have seen in this book are straightforward and elementary programs that will carry out the key functions, most of them with very good accuracy and reasonable speed, but for efficiency the programs "hardwired" into your calculator chips are in many cases quite different and quite a bit more complicated.

Inside Your Calculator: From Simple Programs to Significant Insights   By Gerald R. Rising
Copyright © 2007 John Wiley & Sons, Inc.

How, then, does your calculator really work?[1]

To evaluate trigonometric, logarithmic, exponential, and other functions, many calculators employ a single program represented by the acronym CORDIC for *co*ordinate *r*otation *di*gital *c*omputer. In this chapter I will introduce this approach by sketching a CORDIC-type program that will calculate values for cosine. To do even that will require a review of some additional precalculus mathematics and some additional calculator features. By the end of this chapter you should see why we have used simpler algorithms to accomplish these tasks.

COMPLEX NUMBERS

We first need to review some concepts associated with complex numbers.

Donald Stover has this to say about them in his textbook, *Precalculus Problems and Projects*:

> In their second year of elementary algebra, students typically encounter complex numbers $a + bi$, where a and b are real numbers and i is a so-called "imaginary number" with the mysterious property $i^2 = -1$. Students may find this strange, even illogical. They can visualize a and b as locations on the number line, yet i is nowhere to be seen; they are expected to accept the idea of multiplying this ghostly quantity by b, adding the result to a, and taking the whole thing seriously.

A contributing problem here is the choice of the word "imaginary" to describe those bi numbers. There is a sense in which all numbers are imaginary—they represent abstract human concepts. In this regard consider just one example: you will try in vain to find an instance of a real measurement in the physical world around us. Why? Because the measurements we are limited to in our world allow us only to make approximations. That does not mean that real numbers do not exist, however. Quite the contrary, real numbers are another very important mathematical concept that gives us power over that physical world. We may not be able to display them exactly, but we need numbers like π and $\sqrt{3}$ to deal with everyday mathematical problems.

So, too, do complex numbers give us power over our world. But for historical reasons we are stuck with that misleading word, "imaginary".[2]

---

[1]One way the keys do not work is the way you might hear conjectured by a teacher of calculus: they do not work by evaluating power series.

[2]What word would do better? Perhaps "abstract" or "ideal" would do. Other possibilities might be "orthogonal" or "perpendicular" numbers that would fit the Argand diagram discussion to follow. All would at least avoid the implication of a kind of magic that "imaginary" implies.

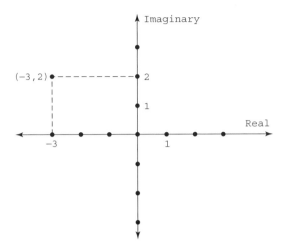

Figure 9.1   The Argand diagram.

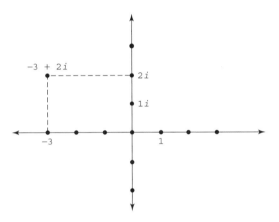

Figure 9.2   The Argand diagram: an alternate view.

A useful and important way of dealing with complex numbers, which are indeed represented by $a + bi$, with a and b real numbers (we'll return to the role of that i shortly), is to extend the number line into a number plane, called by mathematicians an *Argand diagram*. On that plane we plot the point (a, b) to represent $a + bi$. Figure 9.1 shows how the complex number $-3 + 2i$ would be plotted.

On this diagram the x axis is now called the *real axis* and the y axis is called the *imaginary axis*. Note that no i is necessary on the Argand diagram. We can, however, use it by assigning the units on the imaginary axis $1i, 2i, 3i$, and so on, and calling the point $(a, b), a + bi$, as in Figure 9.2. This is simply a matter of notation.

This representation takes much of the magic out of complex numbers. Their imaginary component is simply directed $90°$ from their real component. However, this does not tell the entire story. We need more than a way to represent complex numbers: we want to be able to calculate with them just as we do with rational and real numbers.

The following rules are arbitrary, but they were not chosen by mathematicians simply to confuse us. Rather, they were chosen to make complex numbers conform to requirements imposed on mathematical systems.[3] (In what follows we will continue with standard usage, calling the parts of the complex number $a + bi$; with a the real part and bi the imaginary part.)

We need to define the two operations for complex numbers: addition and multiplication. Subtraction and division depend on those definitions, simply "undoing" them. In defining these operations we must be careful because we will be adding and multiplying the real numbers a, b, c and d in the process of defining the quite different rules for addition and multiplication of complex numbers. To be sure that we keep the two separate, I will temporarily use circled symbols, $\oplus$ and $\circledast$, to indicate complex operations, $+$ and $*$, our familiar real-number operations. Once the operations are defined, however, these distinctions will be dropped.[4]

Addition is straightforward:

$$(a, b) \oplus (c, d) = (a + c, b + d)$$

or

$$(a + bi) \oplus (c + di) = (a + c) + (b + d)i$$

Stated in words, the rule is simply: Add the real coefficients and the imaginary coefficients separately.

Multiplication is much more complicated:

$$(a, b) \circledast (c, d) = (ac - bd, ad + bc)$$

or

$$(a + bi) \circledast (c + di) = (ac - bd) + (ad + bc)i$$

The real coefficient of the product is found by subtracting the product of the imaginary coefficients from the product of the real coefficients of

---

[3] The general properties, which apply to real as well as complex numbers, are called the *field properties*. They include requirements such as the distributive law: that the numbers X, Y and Z obey the rule $X * (Y + Z) = X * Y + X * Z$, for the operations $*$ and $+$ as defined in the system.

[4] Be sure you understand that, although the distinctive signs will then no longer be used, the differences between real and complex operations will remain.

the original numbers. The imaginary coefficient is most easily seen as the "outer" product, ad, plus the "inner" product, bc.

When the operations are defined in this way, consider the product $(0, 1) * (0, 1)$. Here $a = 0, b = 1, c = 0$, and $d = 1$. Substituting these values in the product rule, we have

$$(0, 1) * (0, 1) = (0 * 0 - 1 * 1, 0 * 1 + 1 * 0) = (-1, 0)$$

or

$$(0 + 1i) * (0 + 1i) = -1 + 0i$$

If we prune away those real 0 values from this last statement, we have $i * i = -1$ or $i^2 = -1$, that strange and unexpected result that underlies computation in this system. Taking positive roots of each side of that last equation, we also have the equation that contradicts all that has gone before with real numbers: $i^2 = -1$ is equivalent to $\sqrt{-1} = i$.

Don't misunderstand this. We are not changing the rules for real numbers. Square roots of negative real numbers continue to be disallowed when we deal only with reals. They are allowed only when we work with complex numbers. You can at least justify this to yourself by thinking of these complex numbers as newly defined on that Argand diagram and no longer restricted to a single number line.

There is also a very happy result that these definitions of addition and multiplication of complex numbers (including their attendant, $i^2 = -1$) give us. Those complicated rules for addition and multiplication turn into straightforward algebraic sums and products, and we can work with them in exactly the same way as we work with other algebraic binomials

$$(a + bi) + (c + di) = (a + c) + (b + d)i$$

and

$$(a + bi) * (c + di) = ac + adi + bci + bdi^2$$

but since $i^2 = -1$, this simplifies to $ac + (ad + bd)i - bd$, and further to $(ac - bd) + (ad + bc)i$, which is exactly what our definition tells us to expect.

Numerical examples will show how straightforward these rules are:

$$(3 + 2i) + (5 - 7i) = 8 - 5i$$

and

$$(3 + 2i) * (5 - 7i)$$
$$= 15 + 10i - 21i - 14i^2$$
$$= 15 - 11i - 14(-1)$$
$$= 15 - 11i + 14$$
$$= 29 - 11i$$

I have written out all of the steps in that multiplication example. If you did many such calculations, you would soon omit at least the second and fourth steps.

You can see how this applies by considering how division is accomplished for complex numbers. The result in the form of a definition appears to be a complicated mess:

$$\frac{a + bi}{c + di} = \frac{ac + bd}{c^2 + d^2} + \frac{bc - ad}{c^2 + d^2} i$$

However, we can understand this definition as it relates to an algebraic "trick" we met earlier that is used to rationalize the denominator of fractions containing square roots in the binomial that makes up that denominator.

For example, consider the fraction

$$\frac{2 + \sqrt{3}}{3 - \sqrt{5}}$$

The trick is to multiply this fraction by another carefully chosen fraction whose value is 1, thus giving a product equal to the original fraction. To form this fraction, we choose the *conjugate* of this denominator. This conjugate is the same binomial but with the sign of the second term changed. We use that conjugate, in this case $3 + \sqrt{5}$, in both numerator and denominator of our new factor, thus:

$$\frac{2 + \sqrt{3}}{3 - \sqrt{5}} = \frac{2 + \sqrt{3}}{3 - \sqrt{5}} * 1 = \frac{2 + \sqrt{3}}{3 - \sqrt{5}} * \frac{3 + \sqrt{5}}{3 + \sqrt{5}}$$

This choice of the conjugate now takes advantage of the fact that the product $(a - b)(a + b) = a^2 - b^2$ has no middle term. As if by magic, the radical in the denominator is eliminated (although in this case the numerator becomes more complicated):

$$\frac{2 + \sqrt{3}}{3 - \sqrt{5}} * \frac{3 + \sqrt{5}}{3 + \sqrt{5}} = \frac{6 + 2\sqrt{5} + 3\sqrt{3} + \sqrt{15}}{4}$$

Having recalled how this works, we can apply this algebraic trick to carry out division of complex numbers, because the i with which we are working is a square root and thus will become real when squared. Consider first a numerical example:

$$\frac{2 - 3i}{3 + 5i}$$

We multiply by that carefully chosen fraction formed by use of the conjugate of the original denominator:

$$\frac{2 - 3i}{3 + 5i} = \frac{2 - 3i}{3 + 5i} * \frac{3 - 5i}{3 - 5i}$$

Calculate the algebraic products of numerators and denominators:

$$= \frac{6 - 10i - 9i + 15i^2}{9 - 25i^2}$$

And simplify, using the fact that $i^2 = -1$:

$$= \frac{21 - 19i}{34}$$

Or, separating the real and imaginary parts:

$$= \frac{21}{34} - \frac{19}{34}i$$

Thus, in summary, we have:

$$\frac{2 - 3i}{3 + 5i} = \frac{21}{34} - \frac{19}{34}i$$

(You can check this result against that complicated definition of division by letting $a = 2, b = -3$, and so on and carrying out the computations. You will see that you arrive at the same result.)

Having computed this example, we can follow the same procedure to see the source of that division definition:

$$\frac{a + bi}{c + di} = \frac{a + bi}{c + di} * \frac{c - di}{c - di} = \frac{ac - adi + bci - bdi^2}{c^2 - d^2i^2}$$

$$= \frac{ac + bd + (-ad + bc)i}{c^2 + d^2} = \frac{ac + bd}{c^2 + d^2} + \frac{bc - ad}{c^2 + d^2}i$$

This gives us exactly the formula we sought:

$$\frac{a + bi}{c + di} = \frac{ac + bd}{c^2 + d^2} + \frac{bc - ad}{c^2 + d^2}i$$

Do not miss the point of all that processing. What you have seen is that it is *not necessary* to memorize those complex number definitions of addition, subtraction, multiplication, and division. You simply apply standard algebra, only recalling to use when appropriate $i^2 = -1$ and that conjugate trick.

We now turn to what will appear at the outset to be a completely unrelated topic.

POLAR COORDINATES

The coordinates most familiar to us are rectangular coordinates. To locate a point (a, b) in the plane, we introduce two perpendicular axes—real-number lines meeting at 0 on each, that meeting point called the *origin*. The axes are arbitrarily called the x axis and the y axis. It is usual to have the x axis horizontal and the y axis vertical. New perpendiculars are now drawn from the point (a, b) to these axes and the value assigned to a is the number on the x axis, to b the number on the y axis at the feet of those perpendiculars. Figure 9.3 locates the point $(-3, 2)$.

Figure 9.3 should, of course, be familiar to you. It is almost exactly like Figure 9.1, but with one difference. Here the numbers on both axes are real.

This is, however, only one way to locate points. A useful alternate location system involves what are called *polar coordinates*. To locate a point in polar coordinates, we begin with a single ray—in this case a positive number line in the horizontal direction of the positive x axis of rectangular coordinates. A point (r, A) is then located at a distance r from the initial point of the fixed ray, again called the *origin*, with

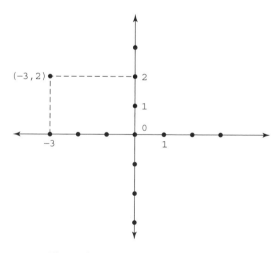

Figure 9.3   Rectangular coordinates.

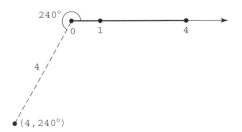

Figure 9.4 Polar coordinates.

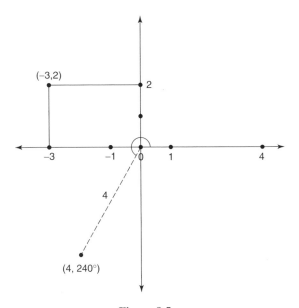

Figure 9.5

its positive angle, *A*, measured counterclockwise from that initial ray. Figure 9.4 locates the point $(4, 240°)$. Notice that alternatively we could use a negative angle here and represent this point as $(4, -120°)$.

We now have two distinct ways to locate points in the plane: rectangular coordinates and polar coordinates. It is very useful to be able to convert from one system to the other. Managing that conversion allows us to utilize whichever system best serves our purposes.

Figure 9.5 shows the points that were introduced in Figures 9.3 and 9.4. We will address the problem of converting each to the other system.

Consider first the point $(4, 240°)$ in polar coordinates. In Figure 9.6, a perpendicular to the negative *y* axis forms a $30°-60°-90°$ triangle whose legs are 2 and $2\sqrt{3}$.

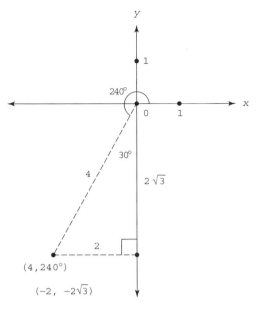

Figure 9.6

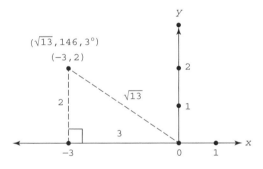

Figure 9.7

Thus $(4, 240°)$ in polar coordinates is equivalent to $(-2, -2\sqrt{3})$ in rectangular coordinates, and, since the degree notation makes clear with which system we are dealing, we can write $(4, 240°) = (-2, -2\sqrt{3})$. Using radian measure, this equation could also be expressed as $(4, 4\pi/3) = (-2, -2\sqrt{3})$, the context making clear which coordinate system we are using.

For the other point $(-3, 2)$, on Figure 9.7 we again form a right triangle, this time by drawing a perpendicular to the x axis. The distance from the origin is the hypotenuse of that triangle and is easily calculated by use

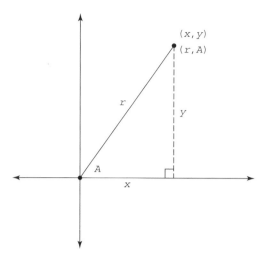

Figure 9.8   Rectangular and polar coordinates.

of the Pythagorean theorem to be $\sqrt{3^2 + 2^2} = \sqrt{13}$. The acute angle in the triangle at the origin is $\cos^{-1}(3/\sqrt{13}) \approx 33.7°.$[5] Therefore the angle measured counterclockwise from the origin is $180° - 33.7° = 146.3°$. Thus we have $(3, -2) \approx (\sqrt{13}, 146.3°).$[6]

We can summarize the relationships between rectangular coordinates (x, y) and polar coordinates (r, A) in the first quadrant as follows (see also Figure 9.8):

Polar to Rectangular	Rectangular to Polar
$x = r \cos A$	$\sqrt{x^2 + y^2} = r$
$y = r \sin A$	$\cos^{-1}(x/r) = A$

Programs for carrying out these conversions are presented next. The program `RECTPOL`[7] which converts from rectangular coordinates, is longer because angle A must be adjusted (through use of S in the program)

---

[5]Recall from the last chapter that $\cos^{-1}$ or arccos is the inverse cosine function. Thus $\cos^{-1}(2/3)$ is the angle whose cosine is 2/3.

[6]The more usual development is to use $\tan^{-1}(y/x) = A$. We adopt this alternate approach by using $\cos^{-1}(x/r)$ because, as you will see in the program we will develop, we avoid the problem that arises when $x = 0$, which would cause y/x to be undefined. When $r = 0$, it is easy in a program simply to report the origin, 0. (No angle is appropriate in that case.)

[7]Many scientific calculators have keys that carry out this and the following `POLRECT` tasks.

to appear in the correct quadrant and to be between $0°$ and $360°$. This program appears first:

```
PROGRAM:RECTPOL
: Prompt X,Y
: 1→S
: If Y < 0
:      ⁻1→S
: √(X²+Y²)→R
: If R ≠ 0
:    Then
:            Disp R
:            S*cos⁻¹(X/R)→A
:            If A < 0
:                   A+360→A
:            Disp A
:    Else
:            Disp 0
: End (If)
```

And here is the simpler program for conversion from polar to rectangular coordinates:

```
PROGRAM:POLRECT
: Prompt R, D
: R*cos D→X
: R*sin D→Y
: Disp X,Y
```

Polar Coordinates for Complex Numbers

Having converted real numbers between rectangular and polar coordinates, we now apply the same concepts to complex numbers. Recalling our coordinate system for complex numbers with the point (a, b) located to represent the number $a + bi$, we simply change the rectangular coordinates of Figure 9.8 to the complex coordinates of Figure 9.9. As a result, we have similar coordinate changes, modified only by replacing x with a and y with b.

Polar to Rectangular	Rectangular to Polar
$r * \cos A = a$	$\sqrt{a^2 + b^2} = r$
$r * \sin A = b$	$\cos^{-1}(a/r) = A$

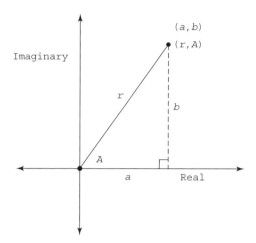

Figure 9.9   Rectangular and polar coordinates for complex numbers.

This seemingly insignificant change, however, carries with it a significant result. We can plug those values for a and b of the polar-to-rectangular conversion into $a + bi$, and we have $a + bi = r * \cos A + (r * \sin A)i$ and, factoring the right side, we have $a + bi = r(\cos A + i * \sin A)$.

Because A appears twice in that second factor, the right side is usually abbreviated

$$r(\cos A + i * \sin A) = r \text{ cis } A \tag{I}$$

and we have $a + bi = r \text{ cis } A$. The right side of that equation is read "r sis A."

While the r cis A abbreviation is useful, it is always important to recall what the abbreviation represents as recorded in equation (I).

Now comes the payoff we seek from this representation.[8] Polar representation provides a straightforward means of carrying out multiplication. We begin with the product of two points expressed in polar form: $(r \text{ cis } A) * (s \text{ cis } B)$.

To carry out the multiplication, we must first replace those abbreviations to give us $r(\cos A + i \sin A) * s(\cos B + i \sin B)$. Reordering the factors produces $rs(\cos A + i \sin A)(\cos B + i \sin B)$.

---

[8]There are other significant payoffs as well. For example, by methods not described here, polar representation of complex numbers allows us to arrive at such results as $\sqrt[4]{-8 - 8i\sqrt{3}} = 1 + i\sqrt{3}, -1 - i\sqrt{3}, \sqrt{3} - i,$ and $-\sqrt{3} + i$; and $e^{i\pi} + 1 = 0$.

Now we multiply the factors in parentheses:

$$rs(\cos A \cos B + i \sin A \cos B + i \cos A \sin B + i^2 \sin A \sin B)$$

We know that $i^2 = -1$, so we can represent the real and imaginary terms in the following way:

$$rs((\cos A \cos B - \sin A \sin B) + i(\sin A \cos B + \cos A \sin B)) \qquad \text{(II)}$$

Here we recall two trigonometric formulas:[9]

$$\cos(A + B) = \cos A \cos B - \sin A \sin B$$

$$\sin(A + B) = \sin A \cos B + \cos A \sin B$$

We substitute those values in equation (II) to obtain $rs(\cos(A + B) + i \sin(A + B))$, and this may be abbreviated rs cis $(A + B)$.

This series of steps establishes the following very important formula for multiplying complex numbers in polar form:

$$(r \text{ cis } A)(s \text{ cis } B) = rs \text{ cis } (A + B)$$

What this formula tells us is that we can multiply the values of two complex numbers expressed in polar form by multiplying their r values[10] and adding their angles. Those points may be considered the arrowheads of two arrows called *vectors* from the origin, so we have a developed a means of multiplying such vectors.

THE CORDIC SIMULATION

Now we have the mathematical tools with which to address the problem of developing a CORDIC simulation program to calculate the cosine of an angle, we'll call it D, entered in degrees.[11]

To accomplish this, we will build up angle D from a series of smaller angles for which trigonometric values are stored in memory. The angles that we will use are (all but the first one approximations; we will see the exact value later) $45°$, $5.7°$, $.57°$, $.057°$, and so on. (I will explain why those particular angles are chosen and how they will be used shortly.)

Suppose we have $D = 63.5°$. We can approach that angle with those given angles as follows:

[9]These formulas are proved in elementary trigonometry texts. We do not offer those proofs here.

[10]The r values are called *moduli* (singular *modulus*). The angles are called *arguments*.

[11]To calculate the cosine of an angle given in radians, you would first convert it to degrees by multiplying the angle by $180/\pi$.

Angle Used	Angles Remaining
45°	63.5°−45° = 18.5°
5.7°	18.5°−5.7° = 12.8°
5.7°	12.8°−5.7° = 7.1°
5.7°	7.1°−5.7° = 1.4°
.57°	1.4°−.57° = .83°
.57°	.83°−.57° = .26°
.057°	.26°−.057° = .203°
.057°	.203−.057° = .146°
.057°	.146°−.057° = .089°
.057°	.089°−.057° = .032°
.0057°	and so on ...

Notice that $45°$ was subtracted once, $5.7°$ three times, $.57°$ twice, and $.057°$ four times, each time approaching the given angle until another subtraction would pass it. This process would continue for a series of smaller and smaller angles, always coming nearer to the given $63.5°$. (At this stage, for example, we are already within $.032°$.)

Now, where did those angles come from? They are angles whose tangents are successively 1, .1, .01, .001, and so on:

$$\tan^{-1}1 = 45°$$

$$\tan^{-1}.1 = 5.710593137°$$

$$\tan^{-1}.01 = .5729386977°$$

$$\tan^{-1}.001 = .0572957604°$$

$$\tan^{-1}.0001 = .0057295779°$$

That still doesn't explain much, however. It is reasonable to ask why we are dealing with the tangent function here when the angle multiplication we have introduced involves sines and cosines. There is a reason. If we used sines and cosines, we would have to carry two functions from step to step. We get around this through use of the basic trigonometric identity:

$$\tan A = \frac{\sin A}{\cos A}$$

Recall our polar point representation: $r(\cos A + i \sin A)$. We multiply this expression by $\cos A/\cos A$, multiplying r by the numerator and dividing the value in parentheses by the denominator. This gives us:

$$r \cos A * \left(\frac{\cos A}{\cos A} + i \frac{\sin A}{\cos A} \right)$$

which simplifies to $r \cos A * (1 + i \tan A)$. Since we're concerned here only with the angle, we replace $r \cos A$ with R to give us $R(1 + i \tan A)$. That series of steps establishes that, when we are concerned only with angles, we can represent $\cos A + i \sin A$ by $1 + i \tan A$.

Okay, so now we have a new way to represent a rotation around the origin of A degrees. In other words, so long as we are not concerned about the value of R, multiplying by $1 + i \tan A$ turns a ray through any point A degrees counterclockwise around the origin.

This works for points expressed as complex numbers in rectangular form as well as in polar form.[12] Consider, for example, the point $(X, Y) = X + Yi$. What would be the result of turning this point A° counterclockwise around the origin? We form the product $(1 + i \tan A)(X + Yi)$ and simplify the product with the following steps:

$1 * X + Yi + iX \tan A + i^2 Y \tan A$

$X + i(Y + X \tan A) - Y \tan A$

$(X - Y \tan A) + i(Y + X \tan A)$

and this represents the point

$(X - Y \tan A, Y + X \tan A)$

Here is what this means:

Original Point	Turn	Result
(X, Y)	A°	$(X - Y \tan A, Y + X \tan A)$

If we now let $\tan A = T$, we can also write this as

Original Point	Turn	Result
(X, Y)	A°	$(X - Y * T, Y + X * T)$

Since $\tan A = T$, $A = \tan^{-1} T$ and, as we have noted, we will use the values $T = 1, .1, .01$, and so on in the program, knowing that they correspond to the successive angles $A = \tan^{-1} 1 = 45^\circ$, $\tan^{-1} .1 = 5.710593137^\circ$, $\tan^{-1} .01 = .5729386977^\circ$, and so forth.

---

[12] Alert readers may question this, because I have not established that this kind of "mixed" multiplication is acceptable. It does work; I simply do not complicate matters further justifying this claim.

Now we are almost ready to present the CORDSIM program. Before we do so, however, we need to store those values of angle A in the computer. We will do so in a matrix by use of the following program:[13]

```
PROGRAM:CORDMAT
: {1,13}→dim([A])
: 1→T
: For (N,1,13)
:      tan⁻¹(T)→[A](1,N)
:      .1*T→T
: End (For)
```

Once that program has been run, your calculator will have

$$[A](1, 1) = \tan^{-1} 1 = 45$$

$$[A](1, 2) = \tan^{-1} .1 = 5.710593137$$

$$[A](1, 3) = \tan^{-1} .01 = .5729386977$$

$$[A](1, 4) = \tan^{-1} .001 = .0572957604$$

$$\cdots$$

$$[A](1, 11) = \tan^{-1} .0000000001 = .000000005729577951$$

$$[A](1, 12) = \tan^{-1} .00000000001 = .0000000000572957795$$

$$[A](1, 13) = \tan^{-1} .000000000001 = .00000000000572957795$$

Now we are finally ready for the CORDSIM program:

```
PROGRAM:CORDSIM
 1 : Prompt D
 2 : 1→T
 3 : 1→X
 4 : 0→Y
 5 : For (N,1,13)
 6 :      [A](1,N)→A
 7 :      D−A→D
 8 :      While D ≥ 0
 9 :           X→K
10 :           X−T*Y→X
11 :           Y+T*K→Y
12 :           D−A→D
13 :      End (While)
```

[13]See Appendix A to see how to set up and use a matrix to store values.

```
14  :        D+A→D
15  :        .1*T→T
16  : End (For)
```
17 : Disp X/$\sqrt{(X^2 + Y^2)}$

We'll examine this CORDIC-type program in detail, continuing with our earlier example, cos 63.5°:

1. : `Prompt D` This program calculates the cosine of the angle D in degrees entered here. Within the program D will represent the angle remaining as it is reduced. For our example 63.5 would be entered here.

2. : `1→T` T represents the tangent of the current angle (A) to be subtracted. After line 2, $T = 1$, which is $\tan 45°$, with $45°$ the first angle we will be subtracting.

3. : `1→X`

4. : `0→Y` These two lines set the initial coordinates of the point (X, Y). Here they are $(X, Y) = (1, 0) = 1 + 0i$. This is a point on the positive X axis, the initial side of the angle whose cosine we seek. Figure 9.10 depicts our situation for our example at this point

5. : `For (N,1,13)` All but the final line of the program will be within this `For` loop. It will run from 1 to 13 because we have stored 13 values of angle A in the matrix $[A]$. Each time through this `For` loop another angle will be processed.

6. : `[A](1,N)→A` The current angle A is retrieved from matrix $[A]$. The first time through the `For` loop $[A](1, 1) = 45$, the angle whose tangent, 1, was set in line 2.

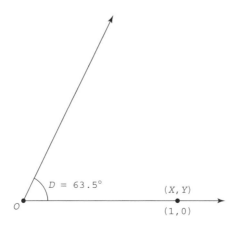

Figure 9.10

7. : D−A→D Now we attempt our first subtraction. In our example this is $62.5 - 45 = 17.5$, so D has now been reduced to 17.5.

8. : While D≥0 This While loop represents the heart of the program. First, notice that the control of this loop is based on D being positive. As soon as it is negative, the loop will be terminated. In other words, as soon as we have subtracted angle A too many times, we will leave the loop. In particular, suppose that our original angle A had been less than 45°. Then, subtracting 45° from it would have given us a negative value. As a result, this While loop would have been skipped and the angle returned to its original value in line 14.

Now we need to see what is going on within that While loop when the angle can be subtracted.

9. : X→K

10. : X − T*Y→X

11. : Y + T*K→Y These are the lines that rotate the vector. They apply the processing we developed earlier:

Original Point	Turn	Result
(X, Y)	A°	$(X - Y * T, Y + X * T)$

There is a problem here, however. If we used just the two instructions, X − T*Y→X and Y + T*X→Y , to carry this out, the X in line 11 would be the new X calculated in line 10 and not the old one that we should be using. That is the reason for introducing the temporary value $X = K$ in line 9.

The result of these lines for our example is shown in Figure 9.11.

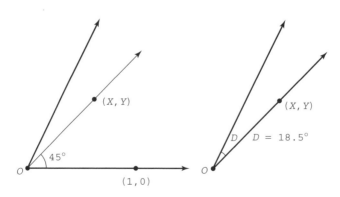

Figure 9.11

The final line in the `While` loop is

12.   : D−A→D This carries out the next subtraction. In our example this line would produce $18.5 - 45 = -26.5$, and we would be left with a negative angle indicating that we have passed the angle whose cosine we seek. This negative value of D pops us out of the `While` loop, and we apply the next instruction.

13.   : End (While)

14.   : D+A→D This line returns the value we subtracted in line 12 just before we exited the `While` loop. In our example, this line would have calculated $-26.5 + 45 = 18.5$, and we would proceed with the angle left after processing the $45°$ angle.

Only one line is left in the `For` loop:

15.   : .1*T→T This line readies us for the next pass through the `For` loop. At this point in our example, we are ready to work with the angle remaining to be processed, $D = 18.5°$ and $T = \tan A = .1 * 1 = .1$. This will be the tangent of the new angle A, 5.710593137, which will be retrieved from matrix $[A]$ in the next pass through the `For` loop.

These values will be processed as was $45°$ and its tangent. $A = 5.710593137$ can be subtracted from $D = 18.5$ three times, leaving a positive value of D, 1.368220589. Each of those values would be processed in the `While` loop. At the end of that third pass, $D - A = -4.342372548$, the loop would be completed, and the value corrected in the following step back to $D = 1.368220589$.

In the same way smaller and smaller angles would be passed through the 13 `For` loops. At the end of all this processing, the point (X, Y) would be very close to the terminal side of the angle $63.5°$.

Figure 9.12 illustrates our situation graphically when we complete the `For` loop.

16.   : End (for)

We can use the triangle that we have created in this diagram to report the cosine value. The triangle legs are X and Y, so its hypotenuse, the length of the vector from the origin to the point (X, Y) is $\sqrt{X^2 + Y^2}$. Thus the cosine of the angle at the origin is

$$\cos 63.5° \approx \frac{X}{\sqrt{X^2 + Y^2}} \approx .4461978131$$

and that is the reason for the final program line:

17.   : Disp X/√ (X²+Y²) Notice that we could equally well have displayed any of the other circular functions:

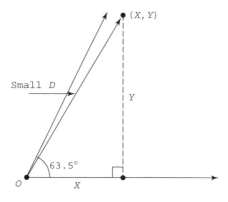

Figure 9.12

Disp Y/X for tangent

Disp X/Y for cotangent

Disp Y/$\sqrt{(X^2 + Y^2)}$ for sine

Disp $\sqrt{(X^2 + Y^2)}$/Y for cosecant

Disp $\sqrt{(X^2 + Y^2)}$/X for secant

I close this chapter with a few comments about this program:

1. Readers should not think that this is *the* CORDIC program; rather, it is an attempt to show some of the ideas used by the engineers who have developed CORDIC.

2. If you run this program, you will see that it is very slow. Clearly, the real CORDIC program that is built into your calculator is far faster. It gives a near-immediate response. Among the reasons the program of this chapter is so much slower is the fact that we had to store values in and retrieve them from a matrix. The actual CORDIC values, on the other hand, are built into calculator memory, and for that reason they are far more readily accessible.

3. What I find interesting and even disturbing about the CORDIC algorithm is that is used to program exponentiation, logarithms, and even square root and multiplication through calculation of trig and even hyperbolic trig functions. Perhaps that is why the extremely simple square root program of Chapter 4 gives you an answer about as quickly as does the CORDIC routine wired into your calculator.

4. For a detailed and carefully developed but intellectually demanding summary of CORDIC programming, see the excellent paper by

Richard Parris, "Elementary Functions and Calculators," available on the web from `math.exeter.edu/rparris/peanut/cordic.pdf` or `www.swarthmore.edu/NatSci/smaurer1/Math6B/Readings/calculatorsParris-rev.pdf`.

PART III

DISPLAYING INFORMATION

10

GRAPHING

> A picture is worth more than ten thousand words.
>
> —Chinese proverb

Graphic calculators offer users the opportunity to picture complex functions and in the process to gain significant insights into their meaning. Consider in this regard just one example. The fishlike collection of trigonometric function graphs in Figure 10.1 is easily drawn and modified.[1]

This chapter is not, however, about drawing complex graphs. It is instead about how you can manage your calculator's screen through simple programs. In doing so, you will gain insight into how electronic engineers program your calculator's graphics keys to perform their tasks. Again we will manage the screen of the TI-84 as an example, but the management of other graphing calculators is similar.

THE CALCULATOR SCREEN

Like your television screen, your calculator screen is made up of hundreds of little squares called pixels (picture elements). Everything that is reported on your screen—numbers, program lines, graphs, and even words—is

---

[1] You can obtain this complex picture on a TI-84 calculator by adjusting your calculator's $\boxed{\text{WINDOW}}$ to $0 \leq X \leq 3.8$ and $^{-}1 \leq Y \leq 1$; changing your angle measure in $\boxed{\text{MODE}}$ to Radian; setting "FORMAT" to AxesOff; keying 12 $\boxed{\text{STO>}}$ N; entering the following three equations in $\boxed{\text{Y=}}$: $Y = \sin(X)$, $Y = ^{-}\sin(X)$ and $Y = \sin(X)\sin(NX)$; and finally keying $\boxed{\text{GRAPH}}$. To modify the striping on the "fish", you can change the value of N by storing a new value in that variable.

Inside Your Calculator: From Simple Programs to Significant Insights   By Gerald R. Rising
Copyright © 2007 John Wiley & Sons, Inc.

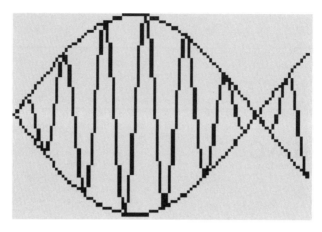

Figure 10.1

shown by darkening (and in some calculators as in computers and TV screens, coloring) those pixels.

The TI-84 screen has 5985 of those pixels, 95 across by 63 down. You can access those pixels by using the "DRAW" `Points` menu. We will begin by using only the command `Pxl-On` from that menu, because that is the tool that parallels what calculator engineers use to build increasingly complex screen images. `Pxl-On` does just what it says: it darkens just one of those 5985 pixels.

Unfortunately, in what a computer scientist friend calls one of the great bloopers of calculator engineering, the pixels are referred to by (*row, column*) and they are numbered from the upper left corner of the screen down and to the right. In the TI-84 the rows are numbered from 0 to 62 down and the columns are numbered 0 to 94 across (see Figure 10.2).[2] This is a near-complete mismatch with our usual (x, y) coordinate system.

Things would be easier if the screen conformed to our normal X, Y orientation with the x axis running to the right and the y axis up. If we locate our origin in the lower left corner, we would like the coordinates to be in the format shown in Figure 10.3.

What this means is that a transformation is always necessary in order to reorient the (X, Y) coordinates in order to place them on the pixel screen. In almost everything you do with pixels, you need to convert from (X, Y)

---

[2]Other screens differ. For example, the CASIO CFX-9850 has 64 rows and 128 columns numbered from (0,0) to (63,127); thus it has 8192 pixels. But it, too, is oriented (*row, column*) and from top to bottom. The TI-92 screen is larger, 105 rows by 240 columns, thus it has 25200 pixels.

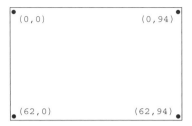

Figure 10.2   Pixel coordinates.

Figure 10.3   First quadrant (X, Y) coordinates.

to (R, C), with R for row and C for column. You can accomplish this by two program lines:[3]

X STO> C

and

62−Y STO> R.

You can use these lines in a program that will allow you to place single pixels on your calculator screen. Here is such a program:

```
PROGRAM:PLOTXY
: Prompt X,Y
: X→C
: 62−Y→R
: Pxl−On(R,C)
```

To place many pixels on your screen, simply use this program over and over. If you wish to erase the screen, use "DRAW" ClrDraw.

Clearly, however, it would take a long time to develop interesting displays by having to run that program to darken each pixel. It turns out that

---

[3]If you are following this discussion with a calculator, you should clear any graph equations you have entered.

we can accomplish what that program does by using the single instruction
`Pxl-On(62−Y,X)` without even writing a program.

What we have accomplished so far is to locate points on the screen as
though the screen represented the first quadrant of a graph. Suppose that
you want instead to plot the points with the origin at the center of the
screen. To do so, you can simply replace two lines of that program

```
: X→C
: 62−Y→R
```

with

```
: 47+X→C
: 31−Y→R
```

or write that single instruction `Pxl−On(31−Y,47+X)`.

AXES AND SCALES

If you drew those graphs at the beginning of this chapter, you used "FOR-
MAT" `AxesOff`. Leaving that setting, you can program what corresponds
to the instruction `AxesOn` yourself, because you can use the `Pxl-On` command to draw those axes. You can also add scales. To carry out these tasks,
clearly you must turn on many pixels, and programming will accomplish
this. `For` loops prove useful here.

In doing this, you don't have to bother with changing (X, Y) to (R, C)
coordinates. You can simply use the original row, column (R, C) format
of the screen.[4] What you want to do is to turn on the pixels in the middle
row and the middle column of the screen. The middle row is the 31st row
(31 lines, 0–30 above; 31 lines, 32–62 below), and the middle column is
the 47th column (47 lines, 0–46 to the left; 47 lines, 48–94 to the right).

Here is a program that will draw these axes:

```
PROGRAM:AXES
: ClrDraw
: For(C,0,94)
:     Pxl−On(31,C)
: End
: For(R,0,62)
:     Pxl−On(R,47)
: End
```

---

[4]This way you also don't have to worry about what coordinates are set in your calculator's
WINDOW.

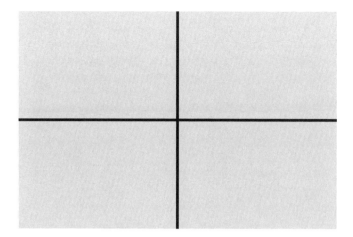

Figure 10.4 AXES output.

That program turns on all 94 pixels across row 31 and all 63 pixels down along column 48. Output of this program is represented in the screen shown in Figure 10.4.

You can now space single pixels along each of these axes to mark off a scale. These pixels are in the row above and therefore one less than that of the X axis (thus the 30th row) and the column to the right and therefore one more than the Y axis (thus the 48th column). You can add two more For loops to the PROGRAM:AXES to do this. Here is the full program:

```
PROGRAM:AXSCALE
: ClrDraw
: For(C,0,94)
:    Pxl−On(31,C)
: End
: For(R,0,62)
:    Pxl−On(R,47)
: End
: For(C,0,18)
:    Pxl−On(30,2+5C)
: End
: For(R,0,12)
:    Pxl−On(1+5R,48)
: End
```

That program places a scale point every 5 pixels. Clearly, minor modifications of that program would space the scale differently.

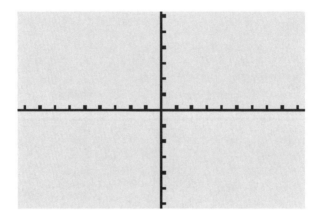

Figure 10.5   AXSCALE output.

BUILDING ON THESE BASICS

Quality programming relies heavily on efficiency. When you work at
the very basic level as you have in the last section, things begin very
*in*efficiently. For example, you would not want to have to add all those
lines of the program AXSCALE (see output graph in Figure 10.5) to every
graph on which you wanted axes and scales to appear. What programmers
do, then, is to make AXSCALE a subroutine to be called up automatically.

Program lines like those of AXSCALE are therefore built into "FORMAT"
AxesOn. The location of those axes, which you could have changed by
modifications of the AXSCALE program, is then controlled by changing
the ranges you choose for X and Y in $\boxed{\text{WINDOW}}$. This menu also allows
modification of the scales on your axes.

You can make graphs appearing on your screen conform to the (X, Y)
pixel coordinates we have been using in the AXSCALE program by setting
"FORMAT" AxesOn and in $\boxed{\text{WINDOW}}$ setting $^-47 \leq X \leq 47$, $^-31 \leq Y \leq$
31, and both X and Y scales to 5.

The command Pt-On builds on the Pxl-On command as well as on the
scale you implemented (which was itself based on the Pxl-On instruction).
When you use the Pt-On command, you are working with regular (X, Y)
coordinates. Just as the axes and scale are built into the AxesOn setting,
the translations from (row, column) that we have shown are built into the
Pt-On instruction by your calculator's internal use of a subroutine.

These represent, of course, only the first steps in this process of building complex structures from simpler basics. But you are seeing in these
development stages what goes on throughout calculator and computer program development. Your word processor, for example, with all its bells

and whistles, is also built up in the same way—by turning on individual screen pixels.

DRAWING GRAPHS

You have seen how that one primitive instruction provides the basis for organizing your screen and plotting points with $X-Y$ coordinates. Now let's see how graphs are drawn. Consider the following program that draws the graph of $Y = X/3$:[5]

```
PROGRAM:LINE
: ClrDraw
: For(X,⁻47,47)
:     X/3→Y
:     Pt−On(X,Y)
: End (For)
```

To run this program as written, you must make the changes in WINDOW and "FORMAT" that were described near the end of the last section.

Now let's see what that program will do. Each time the For loop is run, it produces a point pair that is plotted. It does this rapidly, but you can see the line being formed if you run the program. Here are a few of those pairs:

X	Y
$^-47$	$^-15\frac{2}{3}$
$^-46$	$^-15\frac{1}{3}$
$^-45$	$^-15$
$^-44$	$^-14\frac{2}{3}$
$^-43$	$^-14\frac{1}{3}$

Clearly, those Y values pose a problem. You cannot plot a pixel with coordinates $(^-47,^-15\frac{2}{3})$. The coordinates must have integer values. If you run the program, you will see that the graph isn't the smooth line that you should expect for the linear function, $Y = X/3$. Instead it is a collection of short horizontal segments.

Look closely at those points in Figure 10.6, and you will see that they are rounded to the nearest integer. Thus the plot is really:

---

[5]Leave your WINDOW settings $^-47 \leq X \leq 47$, $^-31 \leq Y \leq 31$ and both X and Y scales to 5. However, you can change "FORMAT" to AxesOn.

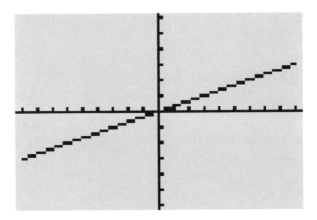

Figure 10.6 $Y = X/3$ graph.

X	Y
$^-47$	$^-16$
$^-46$	$^-15$
$^-45$	$^-15$
$^-44$	$^-15$
$^-43$	$^-14$

You may see how this is working nearer the origin where you have values like:

X	Y
0	0
1	$\frac{1}{3}$
2	$\frac{2}{3}$
3	1
4	$1\frac{1}{3}$
5	$1\frac{2}{3}$

which translate to

X	Y
0	0
1	0
2	1
3	1
4	1
5	2

It should be clear that one program line that is built into the calculator's graphics program is $\boxed{\text{MATH}}$ round. In fact, if you add the step to
PROGRAM:LINE

```
PROGRAM:LINE
: ClrDraw
: For(X,⁻47,47)
:     X/3→Y
:     round(Y)→Y
:     Pt-On(X,Y)
: End (For)
```

you will draw exactly the same graph, having done the calculator's work
for it.

You can demonstrate to yourself that this is what your calculator does
by clearing your graph, using $\boxed{Y=}$ to enter the single equation, $Y = X/3$,
and keying $\boxed{\text{GRAPH}}$. Your resulting graph will appear just like what you
drew with your program.

Your calculator includes some additional smoothing routines. You can
see one of them if you first clear the $Y = X/3$ equation you have entered
in $\boxed{Y=}$, then run the program:

```
PROGRAM:QUAD
: ClrDraw
: For(X,⁻47,47)
:     X²/30−25→Y
:     Pt−On(X,Y)
: End (For)
```

Now go back to $\boxed{Y=}$, enter the function $Y = X^2/30 - 15$, and press
$\boxed{\text{GRAPH}}$. This will give you both parabola graphs on the same screen (see
Figure 10.7) for comparison with your programmed graph below the one
obtained by the calculator routine. Away from the Y axis your graph has
gaps between the plotted pixels. The calculator-produced graph has added
intermediate pixels to make it appear that you have a continuous curve.
When you don't focus on the detail of this screen as we are, the graph
appears continuous. Your program has not done this because it cannot
produce two Y values for the same X. But look closely at the calculator-
produced graph and you will see that it fails the standard vertical line test
for a function when it displays several pixels vertically aligned.[6]

---

[6]This test requires that any function be intersected in at most one point by any vertical
line. It is based on the requirement that, for every X belonging to a function f, its image,
$f(X)$, is unique.

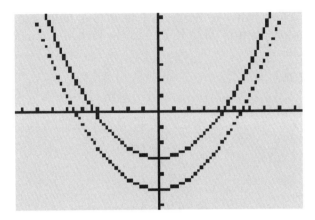

Figure 10.7   Two quadratic graphs.

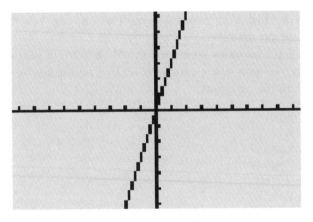

Figure 10.8 $Y = 3X$.

You will see this in even more extreme form if you use your calculator to graph $Y = 3X$ as in Figure 10.8.

You might think that this is simply solved by changing your scales. You can test this by changing the coordinate system in $\boxed{\text{WINDOW}}$. You might try, for example, changing to $^-5 \leq X \leq 5$ and $^-5 \leq Y \leq 5$ with scales both 1. That would give the screen shown in Figure 10.9.

This not only doesn't solve your problem but also distorts your screen. The reason is that the screen is wider than it is high. Look at the scales on the axes, and you will see that they differ. This is something that should always concern you when you are graphing. Since the number of pixels horizontally is 95 and vertically is 63, their ratio is close to 3 : 2. For that reason we might better try $^-6 \leq X \leq 6$ and $^-4 \leq Y \leq 4$.

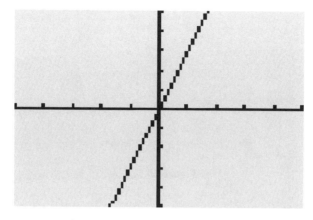

Figure 10.9 $Y = 3X$, scale changed.

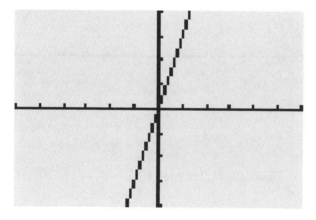

Figure 10.10 $Y = 3X$, scale changed once more.

Surprise! You get exactly the same graph (see Figure 10.10) that you obtained when you first drew $Y = 3X$—except for the scale markings. But think about this. The graph $Y = 3X$ has slope 3 and, as Figure 10.11 shows, that means that $\tan A = 3$ and $A = \tan^{-1} 3$ or $A = 71°$. Thus, so long as you have equally spaced X and Y scales, you will always have this kind of graphing problem at least with this equation. Only "nice" equations like $Y = X$ and $Y = 5$ avoid this kind of problem.

Be sure that you understand what is happening here. Whenever you ask your calculator to graph an equation, your calculator does what you probably did when you first studied graphing using graph paper. Internally it forms a table of corresponding X and Y values and then plots the individual points. Unlike you, however, the calculator constructs that table

Figure 10.11

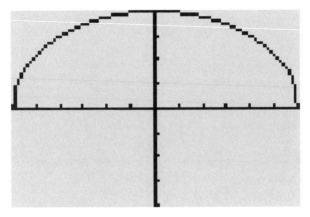

Figure 10.12   The graph of $Y = +2/3\sqrt{36 - X^2}$.

for all 95 X values, no matter what the scale; calculates the corresponding Y values; and plots each (X, Y) point, using rounding in the process to make the points fit on whole-number pixels.

That is why, for example, you cannot ask your calculator to plot the ellipse $4x^2 + 9y^2 = 144$ directly. Instead, you must solve this equation for y to give $y = \pm2/3\sqrt{36 - x^2}$ and then separately plot $Y = +2/3\sqrt{36 - X^2}$ as in Figure 10.12 and separately draw $Y = -2/3\sqrt{36 - X^2}$ as in Figure 10.13 to produce the single graph of the ellipse in Figure 10.14.

Readers should not take these comments as criticisms of the engineers who program these calculators, but rather as insights into the kind of problems they face when dealing with the finite number of pixels on a calculator screen. You are mistaken if you think that these problems are solved simply by having bigger screens with more pixels. The problems are the graphic equivalent of our inability to express irrational numbers

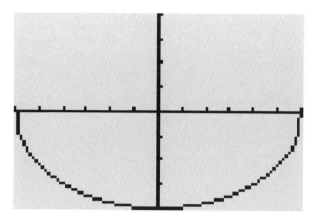

Figure 10.13   The graph of $Y = -2/3\sqrt{36 - X^2}$.

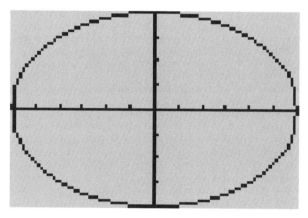

Figure 10.14   The $4X^2 + 9Y^2 = 144$ ellipse.

by terminating decimals. The electronic engineers who devise the wonderful programs that drive calculators and computers have had to adopt
occasional compromises like these.

APPENDIXES

APPENDIX A

A PRIMER ON PROGRAMMING

There are a number of programmable graphing calculators, each with its own idiosyncrasies. The programs in this book are designed for the Texas Instruments TI-84 series and the earlier TI-83 series (see also Figure A.1),[1] all widely used calculators, but they differ in minor ways even from other Texas Instruments calculators like the TI-89 and TI-92.

Despite those minor differences, programming a calculator, including, for example, not only other TI calculators but also calculators such as those of the Casio FX series, is a very straightforward task. Readers of this book who use those other calculators will find their programs to be essentially the same.[2]

What you need to do if you get stuck is look here or in your calculator manual for guidance.

All the following instructions assume that you know how to use your calculator to carry out operations—for example, 3 $*$ 5 $\boxed{\text{ENTER}}$ to multiply 3 by 5, 5 $\boxed{x^2}$ $\boxed{\text{ENTER}}$ to square 5, and $\boxed{\log}$ 5) $\boxed{\text{ENTER}}$ to determine the logarithm of 5.

---

[1]The TI-84 series includes the the TI-84 Plus and the TI-84 Plus Silver Edition. The TI-83 series is similar. This appendix applies to all of these calculators.

[2]Supplements are available for free download to modify this appendix and the programs of this text for other calculators. You may obtain them from www.buffalo.edu/~insrisg/InsideYourCalculator.htm. This website also includes other materials suitable for all readers of this book.

Inside Your Calculator: From Simple Programs to Significant Insights   By Gerald R. Rising
Copyright © 2007 John Wiley & Sons, Inc.

179

Figure A.1   The TI-84 Plus Silver Edition.

USEFUL KEYS

There are several keys that are of general use in working with your calculator. Let's get them out of the way first:

[CLEAR] erases the screen or, when you are editing a program, erases
the current program line.

[DEL] erases the current character or instruction. It also erases a blank
program line.

[(−)] This is a troublesome key. You must be careful to distinguish,
especially in programs, between subtraction, which uses the [−] key, and
negative numbers, which use the [(−)] key. Thus you would write 5 [−] 3,
but [(−)] 3 [+] 5.[3]

The scrolling keys. Near the upper right area of your keyboard is a
group of keys marked with arrowheads or triangles pointing left, right,
up, and down. These keys are useful in moving from line to line and back
and forth within lines on your screen.

[2nd] followed by a key gives you the left-hand instruction above the
key. Three examples: to access the instruction QUIT, press [2nd], then
[MODE]; to write the value of π to your screen, press [2nd], then [∧]; and

---

[3]Standard usage does not make this distinction. Only in programs will this book use the
raised minus sign, as in ⁻3, which is the way such numbers will appear on your calculator
screen, to indicate that this key is to be applied.

to turn your calculator off, press 2nd, then ON. (Don't press two keys at the same time.)

ALPHA is like 2nd, but it gives you the right-hand instruction above the key. These are mostly letters, as you would expect given the name of the key. If you want to type an "A", for example, press ALPHA then MATH. This key works only for the current key, then reverts back to normal operation. To shift to this form for several keys in succession, press 2nd, then ALPHA for A-LOCK; then to shift back, press ALPHA.

MODE gives you access to the way your calculator is formatted. I suggest that you leave all but one key in the leftmost column darkened. The exception is Degree. To make a change, scroll to the desired key and press ENTER. When you are finished with this screen, press "QUIT" (2nd, then MODE).

(*Notation*: From now on I will write a key value that is obtained by 2nd or ALPHA in quotes and no longer give the keys to obtain that value. Thus in the remainder of this appendix I will write "A" and not provide the ALPHA-then-MATH-key route to obtain that value. Those boxes and quotes found in the text will not appear in the actual programs.)

"INS" is a very useful key. In normal entry mode, you type over an entry. For example, if you move your cursor to a 3 and type 2, the 2 will replace the 3. To change to Insert mode, press the "INS" key. Now what you type will be inserted. For example, suppose again that the cursor is on a 3 and you press "INS" 2 7. The 3 would be moved to the right and the 27 would appear in front of it as 273. To add a program line, use "INS" ENTER at the beginning of the line you wish to create. You can exit this mode by typing "INS" again or by moving the cursor.

"QUIT" is also useful as an exit or escape. Unfortunately it will take you out of a program that you are editing. If you use it, then, you must reaccess the program to continue. This instruction does not erase anything and it does not turn your calculator off.

Now finally we're ready to enter and run programs.

A FEW BASIC PROCEDURES

How to Begin Entering a Program

1. Press PRGM.
2. Scroll right twice. That will darken the word NEW on your screen and show below it 1:Create New.
3. Press ENTER. Now your screen reads PROGRAM and on the next line Name=.

4. Type a series of one to seven letters to designate your program. Your calculator is automatically in A-LOCK mode, so you do not need to press ALPHA. In fact, if you want to include numbers in your program name, press ALPHA to exit that alphabet format.
5. When you have named your program, press ENTER.
6. The screen will now show your program name and below it will display a colon, : .
7. Begin typing your program lines, pressing ENTER each time you finish an instruction.

How to Leave a Program You Have Been Editing

1. Press "QUIT".

How to Run a Program Entered in Your Calculator

1. Press PRGM.
2. Scroll down until you darken the program you wish to run.
3. Press ENTER. Now your screen reads prgmNAME, with NAME replaced by the name of the program you will run.
4. Press ENTER again, and the program will run. If your program asks you to enter a number by displaying X=?, type a number followed by ENTER. If your program stops to display information, to continue, press ENTER.
5. When the program is finished, it will display "Done".

How to Enter Program Lines

1. For many entries, you merely press the appropriate keys and the instruction will appear on the screen. For example, suppose that you want to enter the program line $\sqrt{}$ (3)/2)→A, you would type "$\sqrt{}$" 3)/2 STO> "A".

2. Do not spell out instruction names like While or For. These instructions and many others are found by pressing PRGM again. (*Remember*: You are already in Program mode when you are entering program lines.) Three categories appear: CTL, I/O, and EXEC. CTL allows you to select control structures, which will be described later. Seven of them appear on your screen. To choose one of them, scroll down to it (using the down arrow on that ring of arrows) and press ENTER or simply type the number of that instruction. If you scroll beyond instruction 7 you will find still more, including one used quite often: 8:Pause. If you scroll! right, you

will reach I/O. These are input and output commands. The only ones used in this book are 2:Prompt and 3:Disp. The third group, the EXEC instructions, are not used in this book.

3. Press the $\boxed{\text{MATH}}$ key to find some mathematical functions. In this book we have only used those listed under NUM. They include 1:abs(and 5:int(.

4. An important set of relations is to be found by typing "TEST". The items under TEST include =, <, ≥, and so on. Under LOGIC you will find and and or, which are also useful.

How to Edit a Program

Press $\boxed{\text{PRGM}}$, scroll to Edit and scroll down to the program name, press $\boxed{\text{ENTER}}$.

PROGRAM CONTROL STRUCTURES

Unless otherwise instructed, a program proceeds line by line in the order the lines were entered. Control structures (listed under CTL when you press $\boxed{\text{PRGM}}$ while editing) change this order of program operation in specific ways, or stop program operation. In this book we will use many of these keys.

End As you will see in the following examples, this instruction sends control back to the preceding control structure. Ends are like closing parentheses. When one loop is embedded inside another, that loop's End occurs first.

If This is followed by a test, for example, If X=5. When the test is true (in this case if X does equal 5), then the following single program line is performed. When the test is false, that line is skipped. (No Then or End is necessary in this case.)

If
Then
   <program lines>
End Sometimes you want the If test to govern more than one line of code. In that case write those program lines between Then and End.

```
If
Then
    <program line(s)>
Else
    <program line(s)>
```

End When If is used in this way, the program performs the Then instruction(s) when the test is true; the Else instruction or instructions when the test is false.

For This is a counting instruction. It applies a series of steps one at a time within indicated limits. Here is an example of a For loop:

```
: 0 STO> N
: For (I,1,5)
:     N+I STO> N
: End
```

In this example, before we enter the For loop N is set equal to 0. Here there is only one instruction between For and End, but there could have been many.

The For instruction of this example, For (I,1,5), establishes a variable, I, that will take on the values successively between the numbers that follow. In this case these are $I = 1,2,3,4,5$.[4] Each time one of these values of I is set, the instructions that follow are processed. Thus the For loop here is a short way of processing the following five instructions:

```
:         0 + 1 STO> N     (N is now 1)
:         1 + 2 STO> N     (N is now 3)
:         3 + 3 STO> N     (N is now 6)
:         6 + 4 STO> N     (N is now 10)
:        10 + 5 STO> N     (N is now 15)
```

At this point the program leaves the For loop. Any following statements are processed with $N = 15$.

While is a loop that is processed until the test included in the While statement fails. In other words, the While instruction is saying "While this is true, do the following:." Here is a simple example of a While loop:

[4]It is possible to modify this by including another number in the parentheses following the For. Two examples: in For (N,1,9,2), N would jump 2 each time, performing the loop for $N = 1,3,5,7,9$; in For (N,5,3,⁻1), N would reduce one each time, processing $N = 5,4,3$.

```
: 0 STO> N
: While N < 5
:     N+1 STO> N
: End (While)
```

The instruction in this loop will be processed until N is no longer less than 5. Here are the lines that will be processed:

```
:    0 + 1 STO> N     (N is now 1)
:    1 + 1 STO> N     (N is now 2)
:    2 + 1 STO> N     (N is now 3)
:    3 + 1 STO> N     (N is now 4)
:    4 + 1 STO> N     (N is now 5)
```

After this final step $N = 5$, and it is no longer true that $N < 5$. At this point the loop is exited and any instructions following it are processed.

Lbl allows you to set a target line for a Goto instruction. Include a number in this line as in Lbl 1.

Goto sends program control to the designated Lbl line. For example, the line Goto 1 would send program control to Lbl 1. Although these instructions are sometimes useful, they can also cause trouble, creating an endless loop. In this case you must manually stop program operation by pressing ON, then "Quit".[5]

INPUT AND OUTPUT IN PROGRAMS

It is worth repeating here the instructions used in this book that allow the user to enter and retrieve data.

Prompt is accessed when editing a program by again pressing PRGM, scrolling to I/O, and pressing ENTER. If you want to have the program user enter a number N at some point in the program (usually at or near the beginning), enter the line

```
: Prompt N
```

When the program is run, it will stop at that line and display N=?. The user would then type a value and press ENTER. The program would then continue.

---

[5]Early programmers used Goto instructions extensively. Then a computer scientist named Dykstra severely criticized such use because they created programs that looked like spider webs and were very difficult to interpret and debug. Since then this control structure has been rarely used. There are, however, situations when it best serves the programmer's purpose. Goto is used only a few times in appropriate places in this book.

Disp (also on the $\boxed{\text{PRGM}}$ I/O menu) gives a way to present information. If you want to have the program show the value of N, simply enter the line[6]

```
: Disp N
```

This "output line" is useful for other purposes as well. For example, you can use lines like this at various points in your program as an aid to debugging, which is programmer's lingo for finding and correcting errors. When used in the middle of the program, however, you must remember that the program will not stop unless you add another program line. You should enter

```
: Disp N
: Pause
```

This will display the value of N and stop program operation so that you can look at it. (If you fail to include the Pause, the program will display N but continue operation, often so fast that you will miss the report.) To continue, press $\boxed{\text{ENTER}}$. The instruction Pause is on the $\boxed{\text{PRGM}}$ CTL menu. It is number 8.

MATRICES IN PROGRAMS

How to Include a Matrix in a Program

First, let's understand what a matrix is. It is an array of numbers arranged in rows and columns.[7] For example, you might have the following array:

2	0	1	3
1	0	6	6
1	4	9	2

This is a 3×4 matrix. Arbitrarily, we refer to matrix dimensions and matrix entries in the order horizontal row, then vertical column. If we

---

[6]Once a program has been run, you can also access any value that occurred in that program by typing its name followed by $\boxed{\text{ENTER}}$. You could, for example, have omitted the Disp N line from our example and after the program has been run, type N $\boxed{\text{ENTER}}$. Of course, this gives the final value of N, which may differ from its value at the Disp N location.
[7]Matrices (that's the plural of *matrix*) are very useful mathematical structures that can be manipulated in many ways by well-established rules: by addition and multiplication, for example. In this book we will be working only with single matrices used to store and change value entries. Those matrix applications appear in Chapter 9 and Appendix L.

name this matrix [A], we can refer to individual entries. For example, [A](3,4) is 2, the number in the third row and fourth column.

We can also enter or change entries in a matrix. For example, the program line

: 5 $\boxed{\text{STO>}}$ [A](1,4)

would replace the 3 in that matrix with 5.

We must be careful here, however. To write [A], you do not type "[" "A" "]." Instead you press $\boxed{\text{MATRIX}}$, then scroll down (if necessary) until you reach the name you want, in this case [A] and then press $\boxed{\text{ENTER}}$. Your screen will now display [A] and, if you want to refer to a particular array entry, you enter (row,column) with regular keys.

To work with a matrix, you must first establish how many rows and columns you will use and enter the row and column values. When in normal and not program mode, you would press $\boxed{\text{MATRIX}}$, scroll right to EDIT, scroll down to the matrix you wish to use, and press $\boxed{\text{ENTER}}$. You then can change the number of rows and columns you want and change the numbers in the matrix, pressing $\boxed{\text{ENTER}}$ after each change. Press "QUIT" to finish.

How to Set up a Matrix in a Program

Using matrices in a program mostly follows what has already been described. There is one important difference, however. To create a new matrix within a program, you must proceed differently. To establish the matrix [D] with three rows and four columns, for example, use the curly brackets—entered as $\boxed{\text{2nd}}$ (and $\boxed{\text{2nd}}$). Here is the program line:

: {3,4} $\boxed{\text{STO>}}$ dim([D])

After typing $\boxed{\text{STO>}}$ in that line, press $\boxed{\text{MATRIX}}$, scroll right to MATH, and choose dim(. Then you would again press $\boxed{\text{MATRIX}}$ and choose [D] before closing with the final).

PROGRAM APPEARANCE IN THIS TEXT

In order to clarify the structure of programs in this book, they are presented with minor differences from the format in which those programs will be typed. For example, loop instructions are indented and, as noted above, End instructions have the name of the loop included. Programs that you type include none of these.

APPENDIX B

INTERPOLATION

Interpolation is the process of finding intermediate values. Given two values, you seek a third that lies somewhere between them. Consider a simple example. I live near the Canadian border, so I am often confronted with temperatures in Celsius (formerly called *centigrade* and still abbreviated °C). Familiar from childhood with Fahrenheit (F) temperatures, I find it difficult to "think in Celsius." When, for example, a Canadian weathercaster announces that the high in the North Country will be 16°, I have a problem. If I had a thermometer handy, it would show both temperatures, but I do not carry one around with me.

My solution has been to memorize a few corresponding values. Among them are 0°C = 32°F, 10°C = 50°F, and 20°C = 68°F. My thermometer sketch in Figure B.1 displays my situation. It doesn't show the intermediate values on the Fahrenheit scale. Now if I want to know the Fahrenheit temperature that corresponds to 16°C, I interpolate between these values. As a teacher I would have students write a proportion, but it is just as easy for me to think 16 as being six-tenths of the way from 10 to 20, so I want six-tenths of the way from 50 to 68. Since .6 * 18 is about 11, I add 11 to 50 to give the Fahrenheit temperature of 61°.

Although there are other forms of interpolation, the kind most often used in math and science is called *linear interpolation*. This assumes that

Inside Your Calculator: From Simple Programs to Significant Insights   By Gerald R. Rising
Copyright © 2007 John Wiley & Sons, Inc.

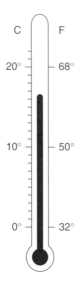

Figure B.1 A thermometer.

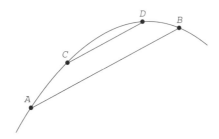

Figure B.2 Linear interpolation on a curve.

the relationship being considered is either linear (as was the temperature example) or "close to" linear.

When we interpolate between trig functions, for example, we know that we're dealing with nonlinear curves. For most curves, linear interpolation between nearby values gives reasonable answers;[1] whereas, as the diagram in Figure B.2 shows, errors increase in size when the distance between values is larger.

For example, in Figure B.2 interpolating along the straight line between points C and D might well produce a satisfactory approximation to the

---

[1]Mathematicians consider a straight line a special kind of curve. For linear curves (to mathematicians "linear" is a synonym for "straight line"), the interpolated value is exact.

curve through those points. But interpolating between points A and B would give results farther from the true values on the curve.

Unfortunately the point of interpolation is lost on those who seek to eliminate it from our schools. The concepts of interpolating to find intermediate values and extrapolating to find values outside the range of known values are important well beyond the bounds of simple computation.

APPENDIX C

PRE–ELECTRONIC CALCULATION TOOLS

Reckoning with numbers is one of the hallmarks of civilization. Throughout history humankind has wanted to know answers to those familiar questions—*how many* and *how much*. In seeking those answers our forebears were led to develop numeration systems and calculating tools.

Quite remarkably, some of our finest early mathematicians worked without the benefit of reasonable systems of numeration. Written in about 300 BCE, Euclid's *Elements*, for example, relies on geometric arguments for the simplest arithmetic operations like addition and multiplication, partly because the numeration systems available to him were so rudimentary.

NUMERATION SYSTEMS AND NUMBER REPRESENTATION

Those early mathematicians had the same built-in tools for calculation that you and I have now: our fingers. Those 10 digits almost certainly provided the motivation for the base 10 or decimal number system in use worldwide today.

Of course, counting by tens is far from the only method of grouping numbers. We find vestiges of other numeration systems in a variety of sources. For example, our time in seconds and minutes and our angle measure in degrees reflect an early number base of 60. And the French

Inside Your Calculator: From Simple Programs to Significant Insights   By Gerald R. Rising
Copyright © 2007 John Wiley & Sons, Inc.

Figure C.1 An abacus.

language, with numbers like *vingt* (20) and *quatre-vingt* (four 20s or 80), displays a number base of 20. Some wags have suggested that early French mathematicians had to remove their shoes in order to carry out their computations.

Likewise, many of our common measurements—12 inches in a foot, 12 hours on a clock dial, 12 months in a year—and our grocery purchases—by dozens and by the dozen dozen or gross—reflect duodecimal (base 12) numeration, a system still encouraged by The Dozenal Societies of Great Britain and America.[1]

Mathematicians paid little attention to alternate numeration systems until the advent of modern computers, when binary numbers provided a remarkably efficient means of associating numbers with electronic circuitry. Until the advent of electronic processing, however, the focus was entirely on the decimal numbers with which you and I compute.

Of the variety of early calculating tools, the abacus (Figure C.1) or counting frame is surely the most familiar. Abaci (also known as *abacuses*) come in several forms. One different from that pictured has one bead above the bar, four beads below. In all forms, however, they remain essentially decimal tools, the rods representing base 10 values reading from the right just as we do in our numeration: units, tens, hundreds, and so on. The lower beads represent units; the upper beads, fives. Thus, the number 158 would be displayed, as on the sketch of Figure C.2, by pushing to the center bar one 5-bead on each of the two rightmost rods, three unit-beads on the farthest right, and one unit-bead on the third bar from the right.

The many rods allow temporary entries to the left while the numbers being calculated are at the right. It takes little experience to become

[1]See "The Dozenal Society," available at www.polar.sunynassau.edu/~dozednal/contact.html.

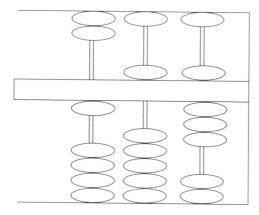

Figure C.2   An abacus representation of 158.

adept at addition and subtraction with this simple tool, but multiplication becomes more complicated and requires both skill and memorization.

We usually associate this device with China, where it was in use by 1200 CE; however, quite similar counting boards were employed by the Babylonians as early as 500 BCE. That all of them were decimally based suggests how universal is the influence of our 10 fingers.

Remarkably, place value notation—units, tens, hundreds, and so on—was not widely adopted for written numbers until about 1500. Before that, numerical calculation (by abacus) and number representation (by clumsy systems like Roman numerals) were separate and oddly unrelated activities.

You can, for example, see some relation between the 158 on the abacus and the Roman numeral representation for that same number, CLVIII, but numbers with 4s or 9s in them are represented very differently; for example, 49 in Roman numerals is XLIX, nothing like the corresponding abacus representation.

NAPIER'S RODS

The abacus was not the only early calculating device designed to short-cut arithmetic processing. The Scottish mathematician John Napier is best known today for his invention of logarithms, but he was far better known during his time—in the early seventeenth century—for his multiplication tools, variously called *Napier's rods* (Figure C.3) or *Napier's bones*, the latter name derived from the fact that expensive sets were often made of bonelike ivory. The popularity of this simple device, widely used by anyone performing calculations, suggests how primitive was the users' arithmetic.

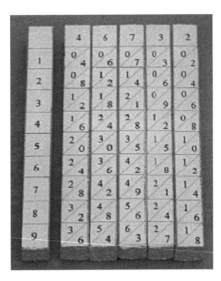

Figure C.3   Napier's rods.

The rods came in two forms, long flat sticks with printing on both sides or rods with square cross section with printing on all four sides. Figure C.3 shows a few of the second type.

Notice that each rod is designed quite simply. A single digit appears in the top square, and beneath it in subsequent squares appear the successive multiples of that digit. For example, under the 2 on the rod at the right appear multiples of 2 with their tens digit and units digit separated by a slanted line: 0/2, 0/4, 0/6, ..., 1/6, 1/8. Placed next to each other, the single digits at the top represent multidigit numbers. When these four rods are placed as in Figure C.4, for example, they support the multiplication of 5276 by various factors.

Suppose, for example, that you wish to multiply 5276 by 57. In Figure C.4 the rows that represent multiples of the digits by 5 and 7 are shaded. It is straightforward to read the products from those rows. They have been separated in Figures C.5 and C.6 to show how they are obtained.

The product digits are read by adding the numbers diagonally. For example, in the 5 row pictured in Figure C.5 there are no carries, but the 7 times row of Figure C.6 has the sum $4 + 9 = 13$ and the 1 is carried to produce $4 + 4 + 1 = 9$ in the next place.

Now the final product 57×5276 is obtained by adding

$$50 \times 5276 = 263800$$

$$\frac{7 \times 5276 = 36932}{57 \times 5276 = 300732}$$

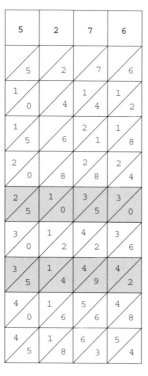

Figure C.4

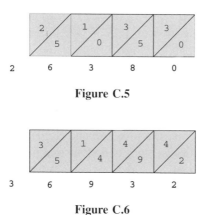

Figure C.5

Figure C.6

It should be clear that the only service this device provides is as a substitute for memorizing the multiples of single-digit numbers. Despite this, the rods were widely used until well into the twentieth century—for over 300 years.

THE FIRST CALCULATING MACHINES

The abacus requires the user to make exchanges between ones and fives and between columns to perform the familiar carrying of addition and borrowing of subtraction. A mechanical calculator carries out the addition process for you. Michael R. Williams has formally defined a mechanical calculator for addition as

> a device that has three properties: a mechanism that will act as a register to store a number; a mechanism to add a fixed amount to the number stored in that register; and an additional mechanism having the ability to deal automatically with any carry, from one digit to the next, that is generated during the addition process.[2]

At least two of these three are apparent to anyone who has used a calculator. If you wish to add 67 and 85, for example, you press keys to store 67, press $+ 85 =$ to add, the calculator performing

$$
\begin{array}{r}
6\ 7 \\
+\ \ \underline{8\ 5} \\
1\ 2 \\
\underline{1\ 4\ 0} \\
1\ 5\ 2
\end{array}
$$

in this case two carries (the one in $7 + 5 = 12$ from the units to the tens column, and the one in $1 + 6 + 8 = 15$ from the tens to the hundreds column) to provide the answer 152. While this seems evident to us as it is displayed in an expanded form emphasizing the units, tens, and hundreds columns with those carried digits in italics, millennia passed before humans produced such a mechanism.

The first apparatus to perform those seemingly simple tasks mechanically was not invented until 1623, when Wilhelm Schickard responded to a request from the astronomer Johann Kepler to manufacture such a device. Although Schickard succeeded in making the calculator, Kepler never got to use it for his calculations because it was destroyed in a fire.

Apparently independently in 1645, a young Frenchman, 21-year-old Blaise Pascal, constructed a mechanism to assist his father, an accountant. This first surviving calculating machine is now called the *Pascaline* (Figure C.7) after its inventor. Pascal went on to become a world-renowned mathematician, physical scientist, and philosopher.

[2]See Michael R. Williams, "Calculating Machines," in *Encyclopedia of Computer Science*, Nature Publishing Group, London, 2000. It is this definition that excludes devices like the abacus and Napier's bones from being considered mechanical calculators.

Figure C.7 The Pascaline.

Mechanisms similar to but much simpler than the Pascaline may occasionally be purchased in school supply stores today. They operate on the same principle as the odometer or mileage indicator of your car or motorcycle. On the Pascaline the wheels on the top of the box represent decimal digits and are turned by the operator using a stylus. Each of those six dials is numbered with the 10 digits, 0 through 9. After the machine is set to zero, an addend is entered by turning each dial the appropriate number of digits. If, for example, you rotated the right-hand dial five clicks and the next dial two clicks, 25 would be displayed in those top squares. The dials snap back to their original positions after they are turned. Now if you wish to add another number, its digits are dialed in the same way. The key feature of the machine is the carry effected when the values in any of the top squares pass from nine to zero.

Among those who also constructed early calculators was Gottfried Wilhelm Leibniz, the German who is credited as coinventor with Isaac Newton of the calculus. In the late seventeenth century he made improvements on the gearing that performed the carries of addition and added a feature that gave the operator the ability to multiply. This multiplication procedure remains a feature of mechanical calculators to this day. How it was designed to carry this out will be explained when you meet the Curta calculator later.

Although theoretically Leibniz' calculator (Figure C.8) represented a major advance, it had, as Stan Augarten[3] so succinctly puts it, "one great drawback, much more serious than its inability to carry or borrow numbers automatically—it didn't work. Leibniz's ambition outran his engineering skill, and the only surviving version of the calculator ... is an inoperative relic."

[3]In *Bit by Bit: An Illustrated History of Computers*, Ticknor & Fields, New York, 1984.

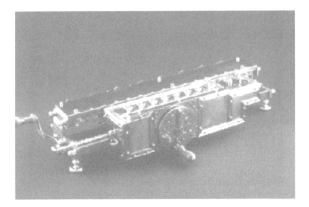

Figure C.8   Leibniz' stepped reckoner.

THE SLIDE RULE

Like the abacus, however, only Leibniz' device offered any even theoretical help with multiplication and division. You could, of course, consider multiplication as multiple addition and calculate simple products like 265×3 by adding $265 + 265 + 265$, but few would wish to calculate 265×23 by adding 23 of those 265s with such primitive machines.

The slide rule (Figure C.9), invented by William Oughtred in 1620—in fact, 3 years before Schickard constructed his adding machine—addresses the problems of multiplication and division, but only by approximation.

A *slide rule* is simply a tool used to add lengths. To see how this works, place two rulers next to each other as shown in Figure C.10. In this case the beginning of the upper ruler (its zero) is placed at the 2 of the lower ruler, and the addition of lengths gives the sums $2 + 0 = 2$, $2 + 1 = 3$, $2 + 2 = 4$, $2 + 3 = 5$, $2 + 4 = 6$, and so on. If intermediate markings were included, fractional values could be added as well. Of course, such calculations as these do not say much for the slide rule. Addition of small numbers like these is hardly a justification for such a device beyond early elementary school.

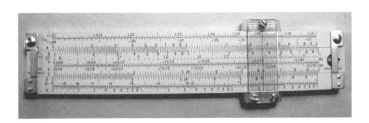

Figure C.9   An engineering slide rule.

Figure C.10 How two rulers add.

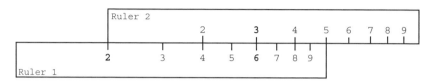

Figure C.11 How a slide rule multiplies.

The power of Oughtred's slide rule resides in the nature of the lengths marked on the ruler scales. He chose markings proportional to the powers of a common base (called *logarithms* by mathematicians—see Chapter 7) to make a slide rule that multiplies.[4]

In Figure C.11 we see a simplified picture of a slide rule aligned to multiply by 2. Unlike the ends in the addition rule, the ends of each rule seen here represent 1 (or a power of 10). In the position displayed we can read from the rule: $2 \cdot 1 = 2$, $2 \cdot 2 = 4$, $2 \cdot 3 = 6$, $2 \cdot 4 = 8$, and $2 \cdot 5 = 10$.[5] Notice that we could then use the rules themselves to help us place additional markings. Thus, for example, directly above the 3 on ruler 1 we could mark 1.5 on ruler 2, since $2 * 1.5 = 3$.

Over time slide rules became more and more complex and were marked with many additional scales. Some scales are, for example, proportional to the values of trigonometric and exponential functions.[6] Other scales allow the user to divide, to square, and to take square roots directly.

---

[4]It is quite easy to construct a multiplying slide rule by reading log values from a calculator or table. For simplicity, assume a rule 10 inches long. Since $\log 2 = .301\ldots$, mark 2 at 3 inches (three-tenths of the 10-inch length) from the left end of the rule. Similarly, since $\log 3 = .477\ldots$, mark 3 at 4.8 inches. Continue by this means to mark the rest of the rule. Two of these marked rules combine to make a slide rule. A way to do this without reference to tables is described in Appendix P.

[5]As well as $20 \cdot 2 = 40$, $20 \cdot 400 = 4000$, $0.2 \cdot 0.004 = 0.0008$, and so on. The user must place the decimal point appropriately by observation or the application of rules learned in elementary school.

[6]Log-log scales—there are usually three of them, marked LL_1, LL_2, and LL_3—give the user direct calculation of x^y for values of x between 1.001 and about 22,000. Some slide rules provide still more log-log scales—marked LL/0, LL/1, LL/2, and LL/3—to allow direct calculation of powers when x is between 0 and 0.999.

There is even a log scale on many slide rules; its units are simply equally spaced, just as on the addition slide rule. For more accurate reading, the scales were very finely marked and a sliding hairline cursor was added. Meanwhile, circular and even cylindrical models appeared, all based on the same principles. Despite these improvements, however, slide rules remained limited to three or occasionally four-digit accuracy.

REFINEMENTS

What is remarkable about this survey is the fact that until World War II, that is essentially the entire story of the mathematics involved with mechanical calculating. For three centuries the only improvements in calculating instruments were refinements of these devices made by a variety of manufacturers, many of them American.

Although little new mathematical processing power was added by these machines, they became increasingly complex, and many of them contained beautifully machined internal parts. An excellent example of this is the Curta universal pocket-size calculating machine that was developed and manufactured in Lichtenstein.

Figure C.12   A Curta calculator.

Although this beautiful instrument is still basically an adding–subtracting machine, by careful manipulation it may also be used to multiply and divide and even, through a very complex process, to take square roots. (The technique is related to the first square root algorithm considered in Chapter 4.)

To add 456 with the Curta, the digits are set by sliding down the keys on the side of the instrument and then the crank is turned. To subtract the same number, the crank is rotated in the opposite direction. To multiply 456 by 3, the crank is rotated three times—thus interpreting multiplication as multiple addition:

$$
\begin{array}{r}
456 \\
456 \\
+\,456 \\
\hline
1368
\end{array}
$$

To multiply your original number by 23, after multiplying by 3 the dial is twisted one place and the crank then rotated two more times, thus mimicking our standard algorithm but with multiple addition at each stage:

$$
\begin{array}{r}
456 \\
\times\ \ 23 \\
\hline
456 \\
456 \\
456 \\
456 \\
+\ 456 \\
\hline
10488
\end{array}
$$

This was exactly the feature first introduced by Leibniz 300 years earlier.) The results of these operations are represented around the edge of the top of the Curta calculator. Rotating the machine with that circular lever then zeroes any results and readies the instrument for a new calculation.

Division on the Curta is carried out by multiple subtraction, an algorithm used today in some European countries and once proposed as a standard method in the United States.[7] The quotient of 10488 divided

---

[7]Employed in some "new math" projects and by the Scott Foresman Company in elementary school arithmetic texts published during the 1960s.

by 456 is equivalent to the number of times 456 may be successively subtracted from 10488 before a remainder of 0 is obtained. Thus, just as multiplication may be interpreted as multiple addition, division may be interpreted as multiple subtraction. You could, of course, simply subtract 456 again and again, keeping track of the number of times you had to subtract it. But you can also shortcut your work just as you did with multiplication, first subtracting 456s ten at a time, thus subtracting each time 456×10 or 4560. Here is this procedure worked out with the number of times subtracted recorded in a separate column to the right:

$$
\begin{array}{rr}
10488 & \\
-\ \ \underline{4560} & 10 \\
5928 & \\
-\ \ \underline{4560} & 10 \\
1368 & \\
-\ \ \underline{456} & 1 \\
912 & \\
-\ \ \underline{456} & 1 \\
456 & \\
-\ \ \underline{456} & \underline{+1} \\
0 & 23 \\
\end{array}
$$

This delicate little Curta calculator indicates how any adding-subtracting machine may be used through careful operation to perform at least multiplication and division as well.

BEHIND THE SCENES

This history brings us up to the 1960s in terms of public information about computers. Although more was going on behind the scenes,[8] in offices around the world, including even industrial and university research laboratories, calculation was carried out by instruments such as those described in this appendix. Discrete calculations, that is, calculations with digits, just reached the level of multiplication and division, and those

---

[8]During the nineteenth and early twentieth centuries, people like George Babbage and Herman Hollerith were setting the stage for the coming computer revolution, but their work was little known to the public.

operations only by mechanically repeating addition and subtraction. More advanced calculations with trig functions, powers, roots, and logarithms were almost always carried out only approximately through the use of slide rules.

And so, the picture of calculating in 1960 we are left with is that of clerks in offices bent over adding machines, and engineering students on university campuses everywhere defined by their slide rules, usually carried in leather holsters hanging from their belts.

But the stage was set for rapid change, partly by developments that had not yet reached the public and partly by theoretical advances. The scientific demands of World War II and the advent of rocketry and the space age provided the impetus for a revolution in calculation. Electronics would fuel that revolution.

APPENDIX D

FERMAT'S LAST THEOREM

To understand Fermat's last theorem or, as it is probably more appropriately known, Fermat's conjecture, recall from school geometry the Pythagorean proposition that holds for legs x and y, and hypotenuse z of any right triangle, $x^2 + y^2 = z^2$.

An infinite number of positive integer solutions fulfill this equation. For example, $x = 3$, $y = 4$, and $z = 5$ satisfy it since $3^2 + 4^2 = 5^2$. Other solutions include $x = 5$, $y = 12$, and $z = 13$; $x = 8$, $y = 15$, and $z = 17$; and $x = 65$, $y = 72$, and $z = 97$. You can easily check these values with a calculator.

In the midseventeenth century, Pierre de Fermat wrote in the margin of an ancient Greek mathematics text that a similar relationship does not hold for higher integer powers, that is, for $x^n + y^n = z^n$, $n \geq 3$; x, y, z, and n positive integers.

For example, we cannot find whole-number triplets like those for the Pythagorean equation, to satisfy equations like $x^3 + y^3 = z^3$.

Fermat also wrote in that margin: "I have discovered a truly marvelous demonstration of this proposition that this margin is too narrow to contain." Some mathematicians think that he wrote this as a tantalizing joke because that "truly marvelous demonstration" or, in fact, any proof was not discovered for over three centuries despite the efforts of many mathematicians.

Inside Your Calculator: From Simple Programs to Significant Insights   By Gerald R. Rising
Copyright © 2007 John Wiley & Sons, Inc.

One way of addressing the problem is to seek to prove the conjecture false simply by displaying a set of four integers (x,y,z, and the power n) that makes the equation correct. Computers were utilized to try such substitutions systematically up to extremely large values, but no solutions were forthcoming. Unfortunately, this does not prove the theorem true; a still larger number might have worked.

Of course, some numbers come close to satisfying Fermat's conjecture. For example, $6^3 + 8^3 = 9^3 - 1$ and $9^3 + 10^3 = 12^3 + 1$.

Two equations have recently been suggested by jokers to prove Fermat's conjecture false.[1]

$$1782^{12} + 1841^{12} = 1922^{12} \quad \text{and} \quad 6107^6 + 8919^6 = 9066^6$$

If you test those on your calculator (for example, calculate $6107^6 + 8919^6$ and compare the result with 9066^6), you will find that the results are the same. The problem is that the answers reported are in scientific notation with many digits not displayed.

It is easy to show that the first equation is false even without doing any calculating. The left side adds an even number and an odd number; thus that sum must be odd. The right side is even. The second equation does not contradict parity (odd-even) laws as an odd plus an odd does give an even sum. A check with a computer providing many-digit accuracy shows that the right side is 453127232314 less than the left, however. Not too close!

Finally, in 1994 Andrew J. Wiles of Princeton University[2] developed a proof of the theorem as Fermat stated it (i.e., that there are no possible solutions), but his proof requires whole areas of mathematics unheard of in Fermat's time. If, indeed, Fermat himself had a "truly marvelous demonstration," no one has yet come upon it.

---

[1] The first of these equations appeared on the television comedy *The Simpsons*, whose writers include several highly trained mathematicians and computer scientists. They later showed still another example: $3987^{12} + 4365^{12} = 4472^{12}$.

[2] Although Professor Wiles received many awards for his work, he never received the Nobel Prize. The simple reason: There is no Nobel Prize for mathematics! Despite this, Wiles' Princeton mathematics colleague, John Forbes Nash, Jr., did win a 1994 Nobel Prize, but his was in economic sciences. The deeply troubled Nash, a paranoid schizophrenic, is the subject of the acclaimed 2001 motion picture, *A Beautiful Mind*, and an even better biography by Sylvia Nasar with the same title, which provided the basis for the movie.

APPENDIX E

AN EXTENSION AND AN APPLICATION OF INTEGER DIVISION

Consider again those program lines for integer division:

```
int(N/D)→Q
```

and

```
N−D*Q→R
```

These program lines not only provide additional processing power but also can give you further insight into how our standard long-division algorithm works.

LONG DIVISION

One of the best things about even the simplest of calculators—those four-bangers that include only the operations addition, subtraction, multiplication, and division—is that they save us from having to do long division. Everyone knows the algorithm; it is the calculation that puts us off. Now you can calculate 10 digits of the quotient of exercises like 7987)21358.07639 or 21358.07639/7978 with no difficulty. You simply key in the dividend or numerator, the division key and the divisor or denominator, and press ENTER. Bingo, up pops that quotient:

Inside Your Calculator: From Simple Programs to Significant Insights   By Gerald R. Rising

2.674104969. Just imagine carrying out that division even to two or three digits with paper and pencil!

You can gain some insight into the long-division process, however, by examining a program that matches it. Here is one that will do this:

```
PROGRAM:LONGDIV
: Prompt N,D
: Lbl 1
:     int(N/D)→Q
:     Disp Q
:     Pause
:     N−D*Q→R
:     R*10→N
: Goto 1
```

If you run that program for those same division exercises, it will provide the quotient one digit at a time. It will report 2, then 6, then 7, and so on. It does the processing in essentially the same way you would do it (except for false trials) with paper and pencil.

In the first pass through the loop, integer division is performed in the int(N/D)→Q line for 21358/7978, producing and reporting the $Q = 2$.

$$\begin{array}{r} 2. \\[-2pt] \overline{} \\[-8pt] 7987)\,\overline{21358.07639} \end{array}$$

Then the remainder, 5384.07639, is calculated:

$$\begin{array}{r} 2. \\[-6pt] 7987)\,\overline{21358.07639} \\ \underline{15974} \\ 5384.07639 \end{array}$$

You probably immediately notice that the whole rest of the dividend is "brought down" by this process. In the long division algorithm you want only the next digit. But you can deal with that one digit simply by multiplying this remainder by 10. Then you will have

$$\begin{array}{r} 2. \\[-6pt] 7987)\,\overline{21358.07639} \\ \underline{15974} \\ 53840.7639 \end{array}$$

and when you rename this as our new quotient, you have the division 53840.7639/7987 to perform in the next pass through the loop. But since this is integer division the quotient of that will be the quotient of 53840/7987, just as in our algorithm.

That next pass through the loop gives us

$$
\begin{array}{r}
2.6 \\
7987{\overline{)}\,21358.07639} \\
\underline{15974} \\
53840.7639 \\
\underline{47922} \\
5818.7639
\end{array}
$$

You again multiply that remainder by 10 to give you the integer division 59187/7987 in the third pass.

That doesn't seem like much since you could have obtained a 10-digit quotient simply by using the division key. It does, however, enable you to do something you could not do before: you can continue the division beyond the 10 digits that the calculator reports. Running this program for this same exercise, for example, would not only produce the quotient you reported by direct division but your process would continue giving digits: 2.67410496932515337 and so on until you press ON "QUIT" to stop program operation.[1]

This simple program has then provided you a way to extend long division to as many digits as you wish.

Producing just one digit at a time is rather slow going, and there is an easy way to give yourself more than one in this process. For example, merely change the seventh line in that program from R*10→N to R*1000→N.

That changes the remainder by moving the decimal place three digits to the right instead of one:

$$
\begin{array}{r}
2. \\
7987{\overline{)}\,21358.07639} \\
\underline{15974} \\
5384076.39
\end{array}
$$

Although that division (5384076/7987) in one step with paper and pencil would—unless you are some kind of calculating savant—be beyond your capability, it poses no problem for your calculator. And the quotient (674) would give you the next three digits of your answer instead of just one.

---

[1]By brute-force continuation of paper-and-pencil long division you could have calculated all these digits as well. You would have affixed additional zeros to the dividend as you did so. The program assumes those added zeros as well.

You can, in fact, use this division to more rapidly accumulate quotient decimals. In the exercises, for example, you would record

   2. 674 104 969 325 153 374 233 128 834 355 828 . . .

Notice also that if multiplying by 1000 gives you three digits at a time, multiplying by 100000 would give you five digits at a time. Sure enough, changing that program line to R*100000→N would give you five quotient digits at a time. For our exercise, you would record

   2. 67410 49693 25153 37423 31288 34355 82822 08588 . . .

Recall, however, that you do have an upper limit of digits to work with here. You must avoid calculating products of more than the number of digits that your calculator can handle. With a 10-digit display, you should limit your multiplications of quotient * divisor to that number. Since in our example we had a four-digit divisor, we should limit ourselves to six-digit partial quotients.

EXPLORING REPEATING DECIMALS

The exercise you considered in the last section simply provided a rather mundane example. There are, however, interesting explorations to be carried out with the program you have been using.

It can be proved that rational numbers are those numbers that may be expressed by repeating decimals. It is worth examining some fractions to see what those decimal representations look like.

Some of them you already know: $\frac{1}{3} = .333333333 \ldots$, $\frac{2}{3} = .666666666 \ldots$, and $\frac{1}{2} = .50000000 \ldots$. You can use the program we developed to examine others. (If you use that program as you left it at the end of the section, you can print out five digits at a time.)

Here are the decimal digits for some fractions that don't come out even (as did $\frac{1}{2}$):

$$\frac{5}{7} = .71428\ 57142\ 85714\ 28571\ 42857 \ldots$$

$$\frac{1}{15} = .06666\ 66666\ 66666\ 66666\ 66666 \ldots{}^2$$

---

[2]When printing out in groups, you have to watch for initial zeros as in the first group here. If your output displays fewer than five digits, you must fill in the first digits with zeros.

$$\frac{9}{13} = .69230\ 76923\ 07692\ 30769\ 23076\ldots$$

$$\frac{9}{23} = .39130\ 43478\ 26086\ 95652\ 17391\ldots$$

Instead of writing out such expressions, it is common to abbreviate them by writing a dot over the first and last digits repeated or by showing a bar over them. You see, then, that you could write the first three of these fractions as

$$\frac{5}{7} = .\dot{7}1428\dot{5}\ (\text{or}\ .\overline{714285})$$

$$\frac{1}{15} = .0\overline{6}$$

$$\frac{9}{13} = .\overline{692307}$$

but what about $\frac{9}{23}$? Does that 39 indicate that it is beginning to repeat? Test it and see.

That and other questions are worth exploring, many of them simply by trying many fractions. Are there fractions with especially long cycles? (*Cycle* is a technical name for the digits repeated.) Is there something interesting about the cycle for $\frac{1}{81}$? What do you notice about how many of each digit appears in the cycle for $\frac{1}{61}$? There is a relation between the digits in the first half of any cycle and those in the second half. What is it? Is there a relation between the number of digits in any cycle and the denominator of the fraction generating it? These and other questions can lead you to some interesting number theory.

APPENDIX F

BINARY ARITHMETIC

Just as we have algorithms for decimal calculations like $235 + 507$ and $235 * 507$, we need similar algorithms for binary calculations like $1101 + 1011$ and 1101×1011.

This takes us back all the way to primary grades when we learned how to add and multiply single-digit numbers. We had to learn what are often called the "facts": $3 + 5 = 8$, $3 \times 5 = 15$, and so on. If we include all the single digits 0 through 9, there are 100 such facts for addition and the same number for multiplication. Here they are organized into Tables F.1 and F.2.

When you think about it even now, those 200 facts were quite a bit to learn. And here is where the great gain in binary processing is to be found. We need only four facts for each of the binary tables, as shown in Tables F.4 and F.5.

In the decimal addition table that $1 + 1$ fact gave us 2 (see Table F.3), but we have to recall that the number corresponding to decimal 2 is the binary 10. Thus the corrected table should be as shown in Table F.4.

It turns out that the multiplication table is even simpler, as seen in Table F.5.

You should immediately notice that, unlike the addition table, this multiplication table (Table F.5) involves only single digits.

Inside Your Calculator: From Simple Programs to Significant Insights   By Gerald R. Rising
Copyright © 2007 John Wiley & Sons, Inc.

TABLE F.1. The Decimal Addition Table

+	0	1	2	3	4	5	6	7	8	9
0	0	1	2	3	4	5	6	7	8	9
1	1	2	3	4	5	6	7	8	9	10
2	2	3	4	5	6	7	8	9	10	11
3	3	4	5	6	7	8	9	10	11	12
4	4	5	6	7	8	9	10	11	12	13
5	5	6	7	8	9	10	11	12	13	14
6	6	7	8	9	10	11	12	13	14	15
7	7	8	9	10	11	12	13	14	15	16
8	8	9	10	11	12	13	14	15	16	17
9	9	10	11	12	13	14	15	16	17	18

TABLE F.2. The Decimal Multiplication Table

*	0	1	2	3	4	5	6	7	8	9
0	0	0	0	0	0	0	0	0	0	0
1	0	1	2	3	4	5	6	7	8	9
2	0	2	4	6	8	10	12	14	16	18
3	0	3	6	9	12	15	18	21	24	27
4	0	4	8	12	16	20	24	28	32	36
5	0	5	10	15	20	25	30	35	40	45
6	0	6	12	18	24	30	36	42	48	54
7	0	7	14	21	28	35	42	49	56	63
8	0	8	16	24	32	40	48	56	64	72
9	0	9	18	27	36	45	54	63	72	81

TABLE F.3. The Binary Addition Table (Almost)

+	0	1
0	0	1
1	1	2

TABLE F.4. The Binary Addition Table

+	0	1
0	0	1
1	1	10

The great power of these two binary arithmetic tables derives, however, not from their ease of recall but from their simple relationships with electronic circuitry. In fact it is this relationship that is the second essential (the first has been said to be binary numeration) of the computer revolution.

**TABLE F.5. The Binary
Multiplication Table**

*	0	1
0	0	0
1	0	1

Figure F.1   An electric circuit.

This book is about algorithms and programs and not about calculator circuitry, but some understanding of how that circuitry works will give you a better sense of what takes place inside your calculator. You will see that the statement, "Calculators calculate simply by converting numbers to binary and working with them internally," may be correct but it hides a great deal in that phrase "working with them internally."

You probably learned in a school science class about two basic kinds of electrical circuits: series and parallel. Combinations of such connections are what process binary arithmetic in your calculator.

At the simplest level we can create a circuit to turn on a lightbulb with the connection of Figure F.1.

Here you have a power source, say, a battery, and wires connecting the battery to a lamp.[1] Hooked up in this way, the light will remain on as long as the battery charge and the bulb life last.

If we insert a switch in this circuit, we can turn the bulb on and off, just as we do a flashlight as shown in Figure F.2.

This gives the following (not very interesting) result:

Switch	Lamp
Off	Off
On	On

[1]You can easily create each of these circuits with a small battery, a lightbulb, and pieces of single strand copper wire. You don't even need to purchase switches, but they make the operation a bit easier. If you use a small battery, you don't have to worry about electric shock.

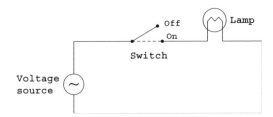

Figure F.2   An electric circuit with a gate.

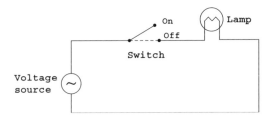

Figure F.3   An electric circuit with a NOT gate.

If we let off be represented by 0 and on by 1, we have the following table:

Switch	Lamp
0	0
1	1

We can play a trick on this table simply by switching the labels on our circuit as in Figure F.3, reversing the labels off and on.

The arrangement shown in Figure F.3 gives us (since we have not switched the status of the bulb)

Switch	Lamp
0	1
1	0

As we will see, this simplistic trick turns out to be very useful. For what should be obvious reasons, it is given the name NOT. It is denoted in electric circuit diagrams as in Figure F.4.

Now we consider the situation with two switches. In the first we have the series circuit of Figure F.5.

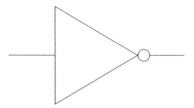

Figure F.4   A NOT gate symbol.

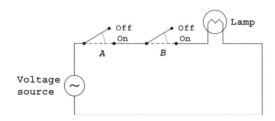

Figure F.5   A series circuit.

Clearly we will only have a circuit when both switches are turned on as in the following scheme:

		Switch B	
		off	on
Switch A	Off	Lamp off	Lamp off
	On	Lamp off	Lamp on

which translates, again with off $= 0$ and on $= 1$, into

		Switch B	
		0	1
Switch A	0	0	0
	1	0	1

This series circuit is called an AND circuit, which is designated in circuit diagrams as in Figure F.6.

The reason it is called an AND circuit or AND gate is its relation to the logic of the word "and" in our grammar. We could have substituted "false" for off and "true" for on and obtained the following scheme:

		B	
AND		False	True
A	False	False	False
	True	False	True

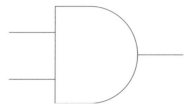

Figure F.6   An AND gate symbol.

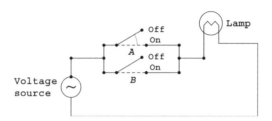

Figure F.7   A parallel circuit.

Consider in this regard the sentence: "If the fair is held today and the weather is good, we will attend." Both "fair is held today" and "weather is good" must be true in order for "we will attend" to be true.

More important, you should recognize that the series table with binary digits is exactly the same as that binary multiplication table. Thus we have an electric circuit, the series circuit, that will, in effect, multiply binary digits. This is a very powerful tool for binary processing.

But now we need a circuit that will add binary digits. It will turn out that this is a good deal more complicated.

When I was a youngster, our strings of Christmas tree lights were hooked up in series; that is, the bulbs were wired one after the other.[2] The problem with this was that, when one bulb went bad or became loose, the entire string of lights went out and much time was taken identifying the bad bulb. To get around this, so-called parallel circuits were soon employed. Figure F.7 is a diagram of a parallel circuit.

On this kind of connection, you need only one of the switches on to have a complete circuit and a lighted lamp. Thus we have

		Switch B	
		off	on
Switch A	Off	Lamp off	Lamp on
	On	Lamp on	Lamp on

[2]The bulbs themselves acted as switches.

or, again with 0 for off and 1 for on:

		Switch B	
		0	1
Switch A	0	0	1
	1	1	1

A parallel circuit is called an OR *gate*, again because of its relation to our use of the word or in grammar.[3] This time the Scheme is

		B	
	OR	False	True
A	False	False	True
	True	True	True

If the weather is good or the fair is held inside, we will attend. The only time you will not attend is when you face both bad weather and an outdoor fair.

A parallel connection is shown in circuit diagrams with the symbol OR of Figure F.8.

Unfortunately, this is not the same as the binary addition table, or we would have both binary multiplication and addition mirrored in simple circuitry.

In fact, reconsider that binary addition table reprised in Table F.6.

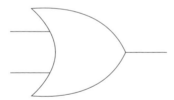

Figure F.8   An OR gate symbol.

**TABLE F.6. The Binary
Addition Table Reprise**

+	0	1
0	0	1
1	1	10

[3]There are two different uses of the word "or" in our language when both phrases are true. The inclusive or that we are using here is the one most commonly used in computer science. An example of the exclusive or is in the phrase: win or lose. You cannot do both at the same time, thus the exclusive or is false when both inputs are 1 or true.

**TABLE F.7. Binary
Addition: Two-Digit Sums**

+	0	1
0	00	01
1	01	10

**TABLE F.8. Binary Addition: Sum and
Carry**

Carry	0	1	Sum	0	1
0	0	0	0	0	1
1	0	1	1	1	0

It should be clear immediately that no circuit would produce that two-digit result of the addition, $1 + 1 = 10$, for our circuits deal only with single digits.

This suggests breaking the binary addition table into two tables. We do this by first making all the sums two-digit, as shown in Table F.7.

We have, of course, not changed the value of the sums by doing this.

Now we make separate tables for the left or carry digit and the right or sum digit, as shown in Table F.8.

Notice immediately that the carry table is like the AND (series) table and binary multiplication, so that is easily represented in circuitry.

The sum table is unlike either the AND gate or the OR gate. It is, however, like a kind of switch pairing that appears in many households. It is called by electricians a "double-pole double-throw switch" but I think of it as the "upstairs-downstairs switch".[4]

Recall how that kind of switch pair works. You come to the foot of the stairs and turn the light on. Then, when you reach the top of the stairs, you switch the light off. Unlike other switches, these are not marked "on" or "off"; rather, they change the status of the circuit from on to off or from off to on.

Accomplishing this task turns out to be a bit complicated. In fact, you need five gates to carry out this seemingly simple task. Figure F.9 is a circuit diagram for the upstairs–downstairs switch.

In the two diagrams of Figure F.10, you can trace the path of the inputs $A = 0$ and $B = 1$ and then $A = 1$ and $B = 1$. You may wish to do the same for $0 + 1$ and $0 + 0$.

---

[4]You may also notice that it is the table that corresponds to the exclusive or of logic mentioned in the previous footnote.

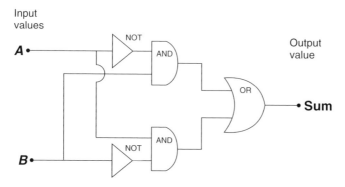

Figure F.9 The upstairs–downstairs switch.

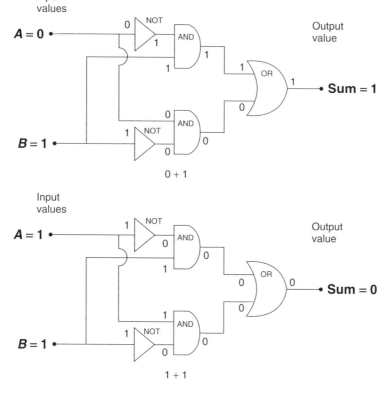

Figure F.10 Upstairs–downstairs switch examples.

Input   Output
values   values

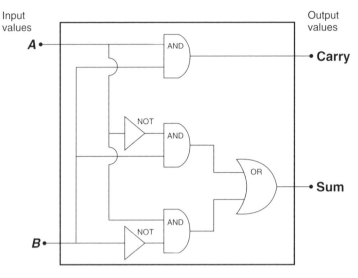

Figure F.11   A half-adder.

All of this circuitry is then combined into what is called a *half-adder* (shown in Figure F.11), which carries out in circuitry single-digit binary addition.

That's quite a complicated process simply to add two single digits, but it turns out that the situation is still worse. Consider, for example, the addition of $11 + 11$. Our half-adder serves us well for the units column. It will provide the sum 10. Following our usual addition algorithm, we record the 0 and carry the 1 to the next column:

$$
\begin{array}{r}
1 \\
11 \\
+11 \\
\hline
0
\end{array}
$$

Now notice that we have the problem of adding three digits, not two! We must find the sum of $1 + 1 + 1$, and our half-adder will not accomplish that. We need what is called a *full-adder*, which takes care of two addends and a carry from the previous column. Figure F.12 shows that complex circuitry.

Figure F.13 shows how $1 + 1 + 1 = 11$ (the third one carried from the previous column) would be processed by a full-adder.

You certainly must have noticed how much more complicated binary addition is than binary multiplication. Binary multiplication took only one

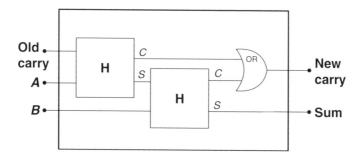

Figure F.12   A full-adder the two Hs represent half-adders.

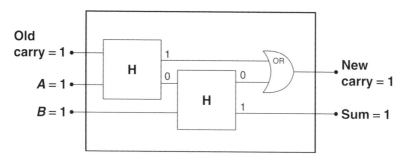

Figure F.13 $1 + 1 + 1 = 11$.

AND gate to accomplish, while binary addition took 6 AND gates (3 in each half-adder), 4 NOT gates (2 in each half-adder), and 3 OR gates (one in each half-adder and one more in the full-adder), for a total of 13 gates. You were probably brought up, as I was, thinking of multiplication (often introduced as multiple addition) as a more complicated and therefore harder process than addition. This reversal should then come as a surprise.

This appendix represents a simplified treatment of some of the electronics necessary to carry out binary arithmetic. You can see some interesting aspects of how subtraction is handled in Appendix G. For a fuller story of the electronics of calculators and computers, I recommend Henry Jacobowitz, *Computer Arithmetic* (Hayden Publishing Company, 1962).

APPENDIX G

BINARY SUBTRACTION

There is an interesting parlor trick that will provide some insight into the way computers subtract binary numbers. With a bit of practice you should be able to use it to entertain your friends.

Tell your audience that you can add nine digit numbers just as fast as a calculator. "Numbers like these," you say, and write down any nine digit number, such as 287,364,338.

Now you invite your audience to give you another nine-digit number. Someone will provide you with a number that you write below the first:[1] If someone offers 482,619,587, write it below your first number.

> 287,364,338
>
> 482,619,587

You add another number to the list:

> 287,364,338
>
> 482,619,587
>
> 517,380,412

[1] If someone suggests a number that is too "regular", like 333,333,333 or 123,456,789, say, "No, those are too easy, anyone can add such numbers; we need tougher, more random numbers." You will see later that there is a reason for this.

Inside Your Calculator: From Simple Programs to Significant Insights   By Gerald R. Rising
Copyright © 2007 John Wiley & Sons, Inc.

"Give me another," you say, and your audience will comply. Now your list reads

$$287,364,338$$
$$482,619,587$$
$$517,380,412$$
$$823,546,734$$

Write a last number below this:

$$287,364,338$$
$$482,619,587$$
$$517,380,412$$
$$823,546,734$$
$$176,453,265$$

Immediately underscore the list, mark a plus sign, and write the answer. Your final result should look like this:

$$287,364,338$$
$$482,619,587$$
$$517,380,412$$
$$823,546,734$$
$$+\ \underline{176,453,265}$$
$$2,287,364,336$$

Of course, the numbers you have written down after the first, were not chosen randomly. When you were given the number, 482,619,587, you wrote what is called the *nines complement* of that number. The nines complement of a digit is the given digit subtracted from nine. For example, the nines complement of 7 is 2. The nines complement of any number is easily written digit by digit. In this case, the pair of numbers are

Audience provides 482,619,587

You write 517,380,412

In the same way, you match that fourth addend:

Audience provides 823,546,734

You write 176,453,265

You should notice the sum of those pairs of numbers. Because you have chosen the nines complement for each digit, in each case the sum will be $999,999,999$, or $1,000,000,000 - 1$.

Your addition exercise then simply converts to

287,364,338	=				287,364,338
482,619,587				+	
517,380,412	=	999,999,999	=		1,000,000,000 − 1
823,546,734				+	
+ 176,453,265	=	999,999,999	=		1,000,000,000 − 1
2,287,364,336		←	←		2,287,364,338 − 2

What you write for the sum is then almost entirely copied from your first number. You begin with 2 billion, but then the rest of the digits except the last are the same as that first addend. Your final digit is simply two 2 less than that first addend's last digit.[2]

We can take this idea of the nines complement in another direction. We can use it in a way that appears to give us a method of subtracting by addition.

Suppose, for example, that we wish to calculate the difference:

$$683,456,125$$
$$- 479,983,512$$

We replace the number being subtracted (the subtrahend) with its nines complement and add

$$683,456,125$$
$$+ 520,016,487$$
$$1,203,472,612$$

Now remove that 1 from the left end of that sum and add it to the units digit. Your result—$203,472,613$—is the answer you seek, the difference between those first numbers.

A bit of algebra will show how this odd method works. We are subtracting two 9-digit numbers: let them be $A - B$. What we have done is to replace B with its nines complement, $999,999,999 - B$, thus we have:

$$A + (999,999,999 - B) = A - B + 999,999,999$$
$$= A - B + 1,000,000,000 - 1$$

---

[2]You can, of course, use more addends, each time after the first matching with a nines complement. Each pair will add another billion and subtract another unit.

To correct for this and obtain the desired $A - B$, we subtract the 1,000,000,000 and add 1.

Now finally we are ready to see how all this applies to binary subtraction.

BINARY SUBTRACTION

As has been pointed out, calculators use binary-coded decimals (BCDs) for computation. The remainder of this appendix applies in ways that are at once more complicated and simpler. We will consider computer *format*, which would have to be modified for calculator application.

Both calculators and computers contain storage registers of fixed length. For example, if a computer calculates with up to 10 decimal digits, it would need registers that would accept 34 binary digits.[3] The reason for this is the largest 10-digit decimal:

$$9,999,999,999_{ten} = 1,001,010,100,000,010,111,110,001,111,111,111_{two}$$

That may seem like a great many binary digits to store, but two things are important to remember here:

1. That 10-digit decimal number was large as well, so you need places for only just over three times as many digits to store that binary number.
2. More important, you gain tremendously in processing capability because each of the binary storage places has only two states, 0 or 1, off or on.

Notice that BCDs actually increase the necessary amount of storage. Recall that you need four binary digits to represent each decimal digit. Thus you would need 40 binary digits to represent a 10-digit number, whereas straight binary representation used only 34.

I remark here that calculators and computers use various means to avoid wasting space when storing small decimal values. For example, a computer would avoid using 31 binary storage digits to store the binary equivalent of 7:

$$0000000000000000000000000000111_{two}$$

Similarly, a calculator would not want to use all 40 binary digits to represent a number like 13 that needs only eight of those digits: 0001 0011.

In any event, binary numbers are stored in registers having a certain number of binary storage units or bits. So that we won't have to write

---

[3]As we have indicated, most 10-digit calculators calculate with one or more additional digits in order to provide more accuracy.

out such lengthy strings to provide examples, I will restrict our registers to eight binary digits or bits. That means that we will confine ourselves to decimal numbers from 0 to 255, because $255_{ten} = 11111111_{two}$.

Now we follow the process we used when we subtracted decimal numbers using the nines complement, except that we use the ones complement. (Just as 9 was the largest decimal digit, 1 is the largest binary digit.) The ones complement is even easier to form. Mathematically it is the binary number subtracted from 11111111, but practically it involves only switching 0s to 1s and 1s to 0s. (Recall from Appendix F that this is easily done digit by digit with single NOT gates.)

Let's see how this works for $213 - 135$. Here is the problem expressed first in decimal and then in binary:

$$
\begin{array}{r}
213 \\
- 135 \\
\hline
\end{array}
\qquad
\begin{array}{r}
11010101 \\
- 10000111 \\
\hline
\end{array}
$$

Now we change this to an addition exercise using the ones complement, given in a third column. I have placed the + in quotes because, as we will see, this is not quite true addition:

$$
\begin{array}{r}
213 \\
- 135 \\
\hline
\end{array}
\qquad
\begin{array}{r}
11010101 \\
- 10000111 \\
\hline
\end{array}
\qquad
\begin{array}{r}
11010101 \\
``+" \; 01111000 \\
\hline
\end{array}
$$

We now add those ones complement binary values:

$$
\begin{array}{r}
11010101 \\
``+" \; 01111000 \\
\hline
101001101
\end{array}
$$

You should notice immediately that there is an extra one at the left end of this answer. We agreed to have only eight places in our register, and that one is in what would be a ninth place. But recall what we did with that one billion in the decimal subtraction with nines complement. We simply remove that one from the left and add it to the rightmost digit:

$$
\begin{array}{r}
11010101 \\
``+" \; 01111000 \\
\hline
101001101 \\
-1 \qquad +1 \\
\hline
01001110
\end{array}
$$

This answer, $1001110_{two} = 78_{ten}$, is just what we want.

Let's see what is happening here. We seek $A - B$, both in binary notation. Instead we calculate $A + (11111111 - B) = A - B + 11111111 = A - B + 100000000 - 1$, so we only have to subtract that 100000000 (throw that leftmost digit away) and add 1 to correct our answer.

Notice that here, unlike decimal subtraction using the nines complement, you simply switch the digits in the subtrahend, add, throw away that extra digit on the left, and add one. Thus the only operation you need to accomplish subtraction is addition.

There is a refinement of this process called *twos complement subtraction* that further simplifies it and avoids even the temporary use of an extra digit of temporary storage.

APPENDIX H

THE RAPID CONVERGENCE OF NEWTON'S METHOD

For the Newton's method square root algorithm, we have proved that $E_{n+1} \leq \frac{1}{2}E_n$, but our example in Chapter 4 showed that the convergence can be much faster than that. For any $N > 1/4$, some further algebraic manipulation will show that, once we have accuracy to n decimal digits, each further step at least doubles the number of accurate digits. More formally, when $E_n \leq 10^{-p}$, $E_{n+1} \leq 10^{-2p}$, for any positive integer p.

Since both sides of the inequality $1/4 < N$ are positive, we can take their square roots to give us $1/2 < \sqrt{N}$, and since we have shown that $\sqrt{N} \leq G_n$ for all $n \geq 1$, we have $1/2 < G_n$. Multiply both sides of this inequality by the positive quantity, $\left(G_n - \sqrt{N}\right)^2 / G_n$, and you arrive at

$$\frac{\left(G_n - \sqrt{N}\right)^2}{2G_n} < \left(G_n - \sqrt{N}\right)^2$$

Notice that the right side of this inequality is E_n^2. Thus we now have

$$\frac{\left(G_n - \sqrt{N}\right)^2}{2G_n} < E_n^2$$

Inside Your Calculator: From Simple Programs to Significant Insights   By Gerald R. Rising
Copyright © 2007 John Wiley & Sons, Inc.

We next want to show that the left side is E_{n+1}. Here is the algebra:

$$\frac{\left(G_n - \sqrt{N}\right)^2}{2G_n} = \frac{G_n{}^2 - 2G_n\sqrt{N} + N}{2G_n}$$

$$= \frac{G_n{}^2 + N}{2G_n} - \frac{2G_n\sqrt{N}}{2G_n} = \frac{G_n + \frac{N}{G_n}}{2} - \sqrt{N}$$

This last is $G_{n+1} - \sqrt{N} = E_{n+1}$. Combining these results, we have shown that $E_{n+1} < E_n{}^2$ is true for $n \geq 1$ and the square root of any $N > 1/4$.

As a consequence, when $E_n \leq 10^{-p}$, for any integer p, then $E_{n+1} < E_n{}^2 \leq (10^{-p})^2 = 10^{-2p}$. Thus the number of digits of accuracy is at least doubled. For example, for $p = 1$, once we have an error less than 0.1, we would have a succession of errors for the following three steps less than 0.01, 0.0001, and 0.00000001, a rapid convergence, indeed.

APPENDIX I

HOW NEWTON'S METHOD APPLIES TO THE SQUARE ROOT ALGORITHM AND THE Rth ROOT OF N

Newton's method provides a far more general procedure than the use we have made of it in finding square roots. For finding a better approximation G_{n+1} of the root of any equation $f(x) = 0$, given an approximation G_n, Newton's method tells us that $G_{n+1} = G_n - [f(G_n)/f'(G_n)]$, with f' representing the derivative of the function f.[1]

Suppose that we seek $\sqrt{N}$ by Newton's method. This is equivalent to solving the equation $x = \sqrt{N}$, which implies $x^2 = N$ and $x^2 - N = 0$. Thus we have $f(x) = x^2 - N$, and its derivative is $f'(x) = 2x$.

Substituting these values in the formula for Newton's method, we have

$$G_{n+1} = G_n - \frac{G_n^2 - N}{2G_n} = \frac{2G_n^2 - G_n^2 + N}{2G_n} = \frac{G_n + \frac{N}{G_n}}{2}$$

In our calculation of square root, we let $N/G_n = H_n$, so this is the same as: $G_{n+1} = (G_n + H_n)/2$.

We have, of course, used this result in our square root program: the arithmetic mean, $(G_n + H_n)/2$, serving as an approximation to $\sqrt{N} = \sqrt{G_n H_n}$, the geometric mean of those same two numbers.

---

[1]The only calculus required in this appendix is the derivitive of x^n, which is nx^{n-1} and the derivative of a constant, N in each instance here, is 0.

Inside Your Calculator: From Simple Programs to Significant Insights   By Gerald R. Rising
Copyright © 2007 John Wiley & Sons, Inc.

It will give us further insight into Newton's method to apply it to the problem of determining cube root. As with square root, we have $x = \sqrt[3]{N}$, $x^3 = N$ and $x^3 - N = 0$. Thus we have $f(x) = x^3 - N$. This time the derivative $f'(x) = 3x^2$.

Substituting these values in the formula for Newton's method, we have

$$G_{n+1} = G_n - \frac{G_n{}^3 - N}{3G_n{}^2} = \frac{3G_n{}^3 - G_n{}^3 + N}{3G_n{}^2} = \frac{2G_n{}^3 + N}{3G_n{}^2} = \frac{2G_n + \frac{N}{G_n{}^2}}{3}$$

We let $N/G_n{}^2 = H_n$, which simplifies our result to:

$$G_{n+1} = \frac{2G_n + H_n}{3}$$

This procedure uses the more general arithmetic mean–geometric mean theorem, which relates n quantities. We will consider it only for $n = 3$, but the extension to more variables should be clear. For $n = 3$, it gives us the following approximation:

$$\frac{a + b + c}{3} \geq \sqrt[3]{abc}$$

In our calculation of cube root we are dealing with three quantities, two G_ns and one H_n, and the arithmetic mean–geometric mean relationship gives us

$$\frac{G_n + G_n + H_n}{3} \geq \sqrt[3]{G_n G_n H_n}$$

Also, since $H_n = N/G_n{}^2$, we have $G_n G_n H_n = N$. Thus we have the approximation

$$\frac{2G_n + H_n}{3} \geq \sqrt[3]{N}$$

We conclude with two more programs—first, a program based on this algorithm that will compute cube root:

```
PROGRAM:CUBERT
: Prompt N
: 10→G
: 0→H
: While G ≠ H
:     N/G²→H
:     (2G+H)/3→G
: End (While)
: Disp G
```

Finally, extending the idea developed for cube root leads to a program to calculate the Rth root of N:

```
PROGRAM:RTHRT
: Prompt N,R
: 10→G
: 0→H
: While G ≠ H
:     N/G^(R−1)→H
:     ((R−1)G+H)/R→G
: End (While)
: Disp G
```

APPENDIX J

THE ANCIENT GREEKS APPROXIMATE $\sqrt{2}$

The Pythagoreans believed that all numbers are rational, that is, that they can be expressed as m/n, with m and n positive integers.[1] It came as a major shock, then, when they discovered that some positive numbers could not be so written.[2] They may have reached this conclusion by the following kind of reasoning.[3]

Beginning with square $ABCD$ with sides n and diagonal m of Figure J.1, they began with the assumption that the ratio m/n is rational, that is, that m and n are positive integers.

On diagonal BD of Figure J.1, construct point F so that $AB = BF$, as in Figure J.2. At point F a perpendicular to BD is then constructed, meeting AD at E. Other segments are constructed to form a square with diagonal ED. The sides of this square are each $m - n$. To determine the length of its diagonal, BE is drawn, forming congruent right triangles ABE and FBE. From these triangles we have $AE = EF = m - n$, and we can determine the diagonal ED by $ED = AD - AE = n - (m - n) = 2n - m$.

---

[1] Zero and negative numbers were unknown to them.

[2] It is said that this upset the Pythagoreans so much that they drowned the discoverer. Perhaps the treatment of scientists today is not so bad after all.

[3] Readers should recall that the Greeks had very poor numeration systems, that they usually represented numbers in terms of the lengths of segments and that they had no algebra. That makes their accomplishments all the more amazing.

Inside Your Calculator: From Simple Programs to Significant Insights   By Gerald R. Rising
Copyright © 2007 John Wiley & Sons, Inc.

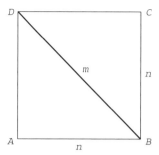

Figure J.1 Diagonal : side ratio in a square.

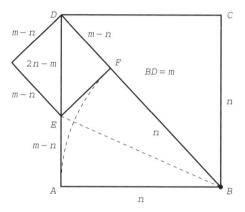

Figure J.2 Diagonal : side ratio on a smaller square.

Thus we have a new square with diagonal : side ratio, $(2n - m)/(m - n)$ and, since all squares are similar, the ratios of their corresponding dimensions are equal and we have

$$\frac{m}{n} = \frac{2n - m}{m - n}$$

There is a problem with that equation. It is straightforward to show that, since the diagonal of the original square is greater than its side, we have $m > n$ and from this, $2m > 2n$ and $m > 2n - m$. We also know that, since the shortest distance between opposite corners of that square is the straight-line diagonal, we have $2n > m$ and $n > m - n$. These two inequalities, $m > 2n - m$ and $n > m - n$, establish that the ratio of diagonal to side in the new square is a fraction with both numerator and denominator smaller than the corresponding values in the original square.

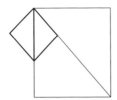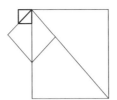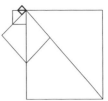

Figure J.3

The problem is that this construction can be repeated over and over as in Figure J.3 with those positive integers getting smaller and smaller. There is a limit to this progression that this procedure contradicts. You cannot have a sequence of positive integers get smaller and smaller forever.[4] This contradiction forces us to conclude that our original premise, that m/n is rational, is false. Thus they were faced with a new kind of number, an irrational number.

We can return to that equation, $m/n = (2n - m)/(m - n)$, and simplify it to obtain successively, $m^2 - mn = 2n^2 - mn$, $m^2 = 2n^2$, and $m = n\sqrt{2}$, as we should expect in modern notation.

Having absorbed the fact that m/n, or to us $\sqrt{2}$, represented a new kind of number, the Pythagoreans set out to explore how m/n might be approximated. They modified the equation $m^2 = 2n^2$ to make it $m^2 \approx 2n^2 \pm 1$. Dividing by n^2 changes this to $m^2/n^2 \approx 2 \pm (1/n^2)$, and we have

$$\frac{m}{n} \approx \sqrt{2 \pm \frac{1}{n^2}}$$

Clearly, increasing the value of n increases the value of n^2 and decreases the value of $1/n^2$. This shows that, for increasing values of n, m/n is a closer and closer approximation to $\sqrt{2}$.

We already know that

$$\frac{m}{n} = \frac{2n - m}{m - n}$$

but this is reducing the values of the numerator and denominator. We want to reverse this process to increase rather than decrease those values. We do this by replacing m by M and n by N in the right-side fraction and, setting numerators and denominators equal, solving for those values. We have $m = 2N - M$ and $n = M - N$, which lead by standard algebraic

---

[4]This limit is stated formally in mathematics as the *well ordering principle*. It states that in any set of positive integers there is a least integer. This simple but remarkably powerful principle is equivalent to mathematical induction.

methods to $M = m + 2n$ and $N = m + n$. As a result of this, we have a new equation, $M/N = (M + 2N)/(M + N)$, with increasing integer values in numerator and denominator, just as we wished.

To use this result, we need only choose seeds (initial values) for M and N. Here is a program that will calculate successive values for M, N, and M/N for any seeds you choose (you should recognize the use of a temporary variable, T, to assist the calculation of new values of M and N in lines 5–7):

```
PROGRAM:SQRT2FR
: Prompt M,N
: Lbl 1
:     Disp M,N,M/N
:     Pause
:     M→K
:     M+2N→M
:     K+N→N
: Goto 1
```

It turns out that any positive integers will serve as seeds.[5] A useful start, however, is provided by setting both equal to 1. Here are the program values obtained with these seeds:

M	N	M/N
1	1	1
3	2	1.5
7	5	1.4
17	12	1.416666667
41	29	1.413793103
99	70	1.414285714
239	169	1.414201183
577	408	1.414215686
1393	985	1.414213198
3363	2378	1.414213625
8119	5741	1.414213552
19601	13860	1.414213564
47321	33461	1.414213562
114243	80782	1.414213562

[5]This is justified in David Flannery's delightful book, *The Square Root of 2: A Dialog Concerning a Number and a Sequence* (Copernicus Books, 2006).

This tells us that we can use the fraction, $\frac{47321}{33461}$, as an estimate of $\sqrt{2}$ equivalent to a 10-digit decimal. Of course, remembering that fraction is no easier than remembering the ten decimal digits.

Using different seeds simply lengthens the processing before converging on $\sqrt{2}$.

It can be shown that a minor change in the program will lead to calculation of $\sqrt{X}$ for any given integer X. For the program line M+2N→M, simply substitute the value of X for 2. For example, to calculate fractions converging on $\sqrt{3}$, the program line would be M+3N→M.

Here, then, is a program that will calculate successive fractions approaching the value of $\sqrt{X}$, for positive integer X, using the arbitrary initial approximation of $\frac{1}{1}$:

```
PROGRAM:SQRTXFR
: Prompt X
: 1→M
: 1→N
: Lbl 1
:    Disp M,N,M/N
:    Pause
:    M→K
:    M+XN→M
:    K+N→N
: Goto 1
```

APPENDIX K

CONTINUED FRACTION APPROXIMATIONS

In Appendix J you saw how to obtain rational fraction approximations to square roots by means of a simple program. Continued fractions provide a way of approximating any positive real number by another kind of ever-improving fractions.

A continued fraction is an expression like this:[1]

$$x = r + \cfrac{1}{s + \cfrac{1}{t + \cfrac{1}{u + \cfrac{1}{v}}}}$$

This expression breaks down a non-negative real number, x, in terms of a series of integers, r, s, t, u, and v (and possibly more following the same pattern), with all but the r positive integers—the integer r can be 0 as well—and all those numerators equal to 1. The digits down the left side of the continued fraction are called *quotients*, and a shorthand notation for this one is a listing of those quotients with a semicolon after the integer. For the example this would be $(r; s, t, u, v)$.

---

[1]Although subscripts are useful in such expressions, I avoid them here in order to simplify the ideas.

Inside Your Calculator: From Simple Programs to Significant Insights   By Gerald R. Rising
Copyright © 2007 John Wiley & Sons, Inc.

Convergents for the same example are r, $r + 1/s$, $r + 1/(s + 1/t)$, and so on. What is important about these convergents is that they do, indeed, get closer and closer to the value of x. In fact, it can be shown that each convergent provides the best fractional estimate of x for denominators of that size or less.

In the case of rational numbers, the continued fraction will always end as in the display shown. Consider, for example, $\frac{13}{29}$. Here is how it would be changed to a continued fraction:

$$\frac{13}{29} = 0 + \frac{1}{\frac{29}{13}} = 0 + \frac{1}{2 + \frac{3}{13}} = 0 + \frac{1}{2 + \frac{1}{\frac{13}{3}}} = 0 + \frac{1}{2 + \frac{1}{4 + \frac{1}{3}}}$$

Thus we have $\frac{13}{29} = (0; 2, 4, 3)$.

Here is how the convergents would approximate $\frac{13}{29}$:

For (0;2) you have $0 + \frac{1}{2} = \frac{1}{2}$.

For (0;2,4) you have

$$0 + \frac{1}{2 + \frac{1}{4}} = 0 + \frac{1}{\frac{9}{4}} = 0 + \frac{4}{9} = \frac{4}{9}$$

And, of course, for (0;2,4,3) you have $\frac{13}{29}$.

That is a great deal of work for little return when dealing with rational numbers, but the idea also applies to irrationals, often with interesting results. Consider, for example, $\sqrt{5}$. Using your calculator to find $\sqrt{5}$ and to determine reciprocals, you would have

$$\sqrt{5} = 2 + .2360679775 = 2 + \frac{1}{4.236067977} = 2 + \frac{1}{4 + .236067977}$$

$$= 2 + \cfrac{1}{4 + \cfrac{1}{4.236067977}} = 2 + \cfrac{1}{4 + \cfrac{1}{4 + \cfrac{1}{4.236067977}}} = \cdots$$

$\cdots$

Notice what is happening here. A pattern has emerged and we have as quotients: $(2;4,4,4,\ldots)$, often displayed as $(2; \overline{4})$.

That this is not merely a chance result of calculation, consider the same process algebraically. We know that $2 < \sqrt{5} < 3$ and that

$$\frac{1}{\sqrt{5} - 2} = \frac{1}{\sqrt{5} - 2} * \frac{\sqrt{5} + 2}{\sqrt{5} + 2} = \frac{\sqrt{5} + 2}{5 - 4} = \sqrt{5} + 2$$

We can use those facts to change $\sqrt{5}$ to a continued fraction:

$$\sqrt{5} = 2 + (\sqrt{5} - 2) = 2 + \cfrac{1}{\cfrac{1}{\sqrt{5} - 2}} = 2 + \cfrac{1}{\sqrt{5} + 2}$$

$$= 2 + \cfrac{1}{4 + (\sqrt{5} - 2)} = 2 + \cfrac{1}{4 + \cfrac{1}{\cfrac{1}{\sqrt{5} - 2}}} = 2 + \cfrac{1}{4 + \cfrac{1}{\sqrt{5} + 2}} = \cdots$$

You should see that both lines end with the same expression to be changed. Thus, more formally, we have once again $\sqrt{5} = (2; \overline{4})$.

While it is interesting that similar patterns occur for other radical expressions, what is more useful is the fact that, as I pointed out earlier, those convergents lead to better and better rational approximations.

Here is a rather complicated program that will calculate for any input value, X, the successive quotients, Q, together with the numerator, A, and denominator, B, of the convergents, and the convergent value.

```
PROGRAM:CONTFRX
: 1→A
: 0→B
: 0→C
: 1→D
: Prompt X
: Lbl 1
:     int(X)→Q
:     A→T
:     QA+C→A          (I)
:     T→C
:     B→T
:     QB+D→B          (II)
:     T→D
:     Disp Q,A,B,A/B
:     Pause
:     (X−int(X))⁻¹→X
: Goto 1
```

This is another program with an infinite loop. To escape, use $\boxed{\text{ON}}$ "QUIT".

I will not attempt to explain this program other than to point out that the key lines are noted as (I) and (II). They process by multiplying the current quotient by the numerator and denominator of two previously reported fractions.

Here are two examples of the use of this program. First, consider $\sqrt{6}$. Note that in the program the quotients are Q, the numerator A, and the denominator B of the convergent:

Pass	Quotient	Numerator	Denominator	Convergent
1	2	2	1	2
2	2	5	2	2.5
3	4	22	9	2.444444444
4	2	49	20	2.45
5	4	218	89	2.449438202
6	2	485	198	2.449494949
7	4	2158	881	2.449489217
8	2	4801	1960	2.449489796
9	4	21362	8721	2.449489737
10	2	47525	19402	2.449489743

and this is $\sqrt{6} \approx \frac{47525}{19402}$, to 10-decimal digit accuracy.

As a second example, consider the results when π is run through this program:

Pass	Quotient	Numerator	Denominator	Convergent
1	3	3	1	3
2	7	22	7	3.142857143
3	15	333	106	3.141509434
4	1	355	113	3.14159292
5	292	103993	33102	3.141592653
6	1	104348	33215	3.141592654

Notice several things about this series of converging fractions. First, you should recognize that the second fraction, $\frac{22}{7} = 3\frac{1}{7}$, is a familiar approximation. But another approximation, which is easy to recall, gives far better accuracy. It is the result of the fourth pass: $\frac{355}{113}$.[2] That value is accurate to six digits, its error a remarkably small .0000085%. Notice how easy it is to remember this fraction, starting from the denominator, you have the sequence 113355, pairs of the first three odd numbers. That fraction is not only much easier to remember than 3.1416 but it is also over 27 times more accurate.

---

[2]That is equivalent to $\pi \approx 3 + \dfrac{1}{7+\frac{1}{16}}$.

APPENDIX L

MULTIPLYING NUMBERS WITH MANY DIGITS

Most calculators display ten digits or less. While it is more than enough for most calculations, this limitation prevents you, for example, from finding directly all the digits in the product of two 6-digit numbers. If you multiply $285694 * 786603$, your calculator will display something like 2.247277575E11, which is equivalent to $2.247277575 * 10^{11}$ or 224727757500, whereas the answer with all of its digits expressed is 224727757482.

More advanced calculators and computers have programs that will display factors and products containing many more digits. In fact, you can use them to obtain the product of 100-digit factors.

What those programs do is carry out algorithms similar to ours for multiplication but with *blocks* of digits instead of individual digits. To see what I mean, compare the following multiplications.

First is the product of 78 and 65 with individual digits.

```
      7   8          7  8            7  8             7   8
  *   6   5       *  6  5         *  6  5          *  6   5
     _____     _____       _____        _____
     35  40         3  9  0         3  9  0          3   9  0
  42  48          4 6  8          4  6  8          4  6   8
  _____     _____       _____        _____
                 4 9 17  0        4 10  7  0        5   0  7  0
```

Inside Your Calculator: From Simple Programs to Significant Insights   By Gerald R. Rising
Copyright © 2007 John Wiley & Sons, Inc.

What I have done here is break down this process into its simple steps. The first step records the products of the digits in their correct locations but without carrying. The second step shows the result after carrying and with the columns added. The last two steps show the result of carrying in the product until the final answer is achieved.

Now I will mirror that processing but with *pairs* of digits in each location. Here we want the product of 6728 and 5692:

$$
\begin{array}{rr}
67 & 28 \\
* \underline{56} & \underline{92} \\
6164 & 2576 \\
\underline{3752 \quad 1568} &
\end{array}
$$

In this first step the four products are formed (by calculator). For example, $92 * 28 = 2576$, and the products are shifted just as in single-digit multiplication.

$$
\begin{array}{rrr}
67 & 28 & \\
* 56 & 92 & \\
61 & 89 & 76 \\
\underline{37 \quad 67 \quad 68} &
\end{array}
$$

Here we have carried the 25 and then the 61 in the first partial product, and the 15 and then the 37 in the second:

$$
\begin{array}{rrrr}
67 & 28 & & \\
* 56 & 92 & & \\
61 & 89 & 76 & \\
\underline{37 \quad 67 \quad 68} & & & \\
37 & 128 & 157 & 76
\end{array}
$$

Here the resulting pairs of digits have been aligned and added:

$$
\begin{array}{rrrr}
67 & 28 & & \\
* 56 & 92 & & \\
61 & 89 & 76 & \\
\underline{37 \quad 67 \quad 68} & & & \\
37 & 128 & 157 & 76
\end{array}
$$

And finally the ones in 128 and 157 are carried:

$$
\begin{array}{rrrr}
 & 67 & 28 \\
* \; 56 & 92 \\
\hline
61 & 89 & 76 \\
37 & 67 & 68 \\
\hline
38 & 29 & 57 & 76 \\
\end{array}
$$

This gives us our final product, 38295776.

Here is a 29-line program to carry out this process for two factors, each with many digits.[1]

```
    PROGRAM:BIGMULT
 1  : Prompt M,N
 2  : M+N−1→T
 3  : {3,T}→dim([A])
 4  : For (I,1,M)
 5  :      Prompt X
 6  :      X→[A](1,I)
 7  : End (For)
 8  : For (I,1,N)
 9  :      Prompt Y
10  :      Y→[A](2,I)
11  : End (For)
12  : For (I,1,T)
13  :      0→[A](3,I)
14  : End (For)
15  : For (J,1,M)
16  :      For(K,1,N)
17  :          J+K−1→I
18  :          [A](3,I)+[A](1,J)*[A](2,K)→[A](3,I)
19  :      End (For)
20  : End (For)
21  : For (I,T,2,⁻1)
22  :      int([A](3,I)/10^4)→C
23  :      [A](3,I)−C*10^4→[A](3,I)
24  :      [A](3,I−1)+C→[A](3,I−1)
25  : End (For)
26  : For (I,1,T)
27  :      Disp [A](3,I)
28  :      Pause
29  : End (For)
```

[1]On the TI-84 calculator the product can have up to 200 digits.

To show how to use this program, I will consider how the product of two large numbers would be processed. If we wanted to find the product of the 17-digit number, 35448796577934447, multiplied by the 15-digit number, 890038376249965, we would first break the factors down into 4-digit groups, working right to left: 35448796577934447 would be 3 5448 7965 7793 4447 and 890038376249965 would be 890 0383 7624 9965.

You now count the number of groups in each factor. The first factor has five groups and the second four. These are the values of M and N requested when you run the program. Once you have entered those two values, you are asked to enter the blocks in each number, left to right. Here is the way the steps would appear when you have correctly entered these factors:

M?	5
N?	4
X?	3
X?	5448
X?	7965
X?	7793
X?	4447
Y?	890
Y?	383
Y?	7624
Y?	9965

If you do that correctly, the program should begin to display the product in four-digit blocks: 3155, 789, 3462, 4009, 1078, 9414, 5604, 4355. (Press ENTER after you record each four-digit block.) Being careful to fill out any block with less than four digits (the 789 becomes 0789) you have the product reported four digits at a time:

3155   0789   3462   4009   1078   9414   5604   4355

I will describe in only general terms how this program works.

In line 3, a matrix is established.[2] It has 3 rows and $M + N - 1$ columns, the number of columns determined by the possible size of the product. In our example with $M = 5$ and $N = 4$, we need a 3×8 matrix, in the program designated [A]. You can think of it as something like this:

___   ___   ___   ___   ___   ___   ___   ___

___   ___   ___   ___   ___   ___   ___   ___

___   ___   ___   ___   ___   ___   ___   ___

[2] Appendix A provides guidelines on how to enter a matrix in a program.

Lines 4–7 enter the first factor groups in the first line and lines 8–11, the second factor groups in the second line:

___3	5448	7965	7793	4447	_____	_____	_____
___890	___383	7624	9965	_____	_____	_____	_____
_____	_____	_____	_____	_____	_____	_____	_____

Lines 12–14 enter 0s in the 3rd line in case values reside there from earlier programs:

___3	5448	7965	7793	4447	_____	_____	_____
___890	___383	7624	9965	_____	_____	_____	_____
___0	___0	___0	___0	___0	___0	___0	___0

Lines 15–20 calculate the partial products and accumulate them in line 3. Notice that these partial products each have up to nine digits. Here are the numbers that would appear in that third row:

$$2670 \quad 4849869 \quad 9198306 \quad 51551812 \quad 121957029$$
$$140488258 \quad 111561173 \quad 44314355$$

Lines 21–25 perform the carrying in order to leave only four digits in each column except the leftmost. Note first that this section of the program works from right to left, the $^{-}1$ in the instruction For(I,T,1,$^{-}$1) indicating that the increment is $^{-}1$ rather than +1.

You can see how this is done by considering how that rightmost entry, 44314355, is processed. First, in line 22 it is divided by 10^4. This gives 4431.4355, and C is the integer value of this number, 4431. In line 23 that number is multiplied by 10^4 to produce 44310000, and that number is subtracted from the original 44314355, leaving the desired Four-digit number, 4355. In line 24, C (still 4431) is added to the next value to the left to change it from 111561173 to 111565604.

Lines 26–29 display the final answer in groups of four digits.

CHECKING YOUR PROGRAM

It is always good procedure to check to see if your programs produce correct results. You could, of course, check the program offered here by using the same numbers that served as an example. But these numbers might have been wrong.

An example like 1,000,000,000 * 100,000,000 is not especially useful because, even if the program did give the correct answer, it would not have involved any carrying and thus would not test parts of the program. Instead, I suggest an example like 999,999,999,999 * 999,999,999,999. This will test (quite severely) those carrying properties, and we can still work out the product for comparison.

999,999,999,999 * 999,999,999,999 is equivalent to

$$(1,000,000,000,000 - 1)^2.$$

We can treat the number in parentheses as a binomial and square it to give:

$$1,000,000,000,000^2 - 2 * 1,000,000,000,000 + 1$$

which is equivalent to

$$1,000,000,000,000,000,000,000,001$$
$$\underline{- \quad 2,000,000,000,000}$$

with the result: 999,999,999,998,000,000,000,001.

This is a nice result to check against your program.

APPENDIX M

FINDING EQUATION ROOTS BY BINARY SEARCH

Many advanced calculators have equation solving routines built in. However, once you know the general location of a root, you can also use a *binary search* program to find most equation roots to considerable accuracy.

We will develop the idea of this process through use of a simple equation, $x^2 - 8x + 13 = 0$. This equation will serve us well because, although its roots are irrational, we can find them to check our computation through use of the quadratic formula.[1] But be sure that you understand that the method will serve equally well for complicated equations involving trig and exponential formulas. For example, you could use it (three times) to approximate to many-digit accuracy the three roots of the equation $1.2 * \ln(x+1) - 2 * \cos^2 x = 0$.

All we need to know in order to use this program are two x values between which the graph of the function $y = x^2 - 8x + 13$ that corresponds to our equation crosses the x axis exactly once. With a graphic calculator, such an interval is easy to find by graphing the function to give a picture like the one in Figure M.1. Clearly, our example crosses the x axis between 2 and 3 and again between 5 and 6. Since the equation of

---

[1]The quadratic formula provides solutions to the general quadratic equation, $ax^2 + bx + c = 0$. It is

$$x = \frac{-b \pm \sqrt{b^2 - 4ac}}{2a}$$

Inside Your Calculator: From Simple Programs to Significant Insights   By Gerald R. Rising
Copyright © 2007 John Wiley & Sons, Inc.

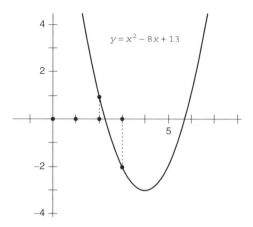

Figure M.1   The graph of $y = x^2 - 8x + 13$.

the x axis is $y = 0$, those crossings represent solutions of the equation $x^2 - 8x + 13 = 0$.

"Crossing the x axis" means that the y value of the function changes from positive to negative or from negative to positive. In both cases the value of the function changes sign. Notice that this means that the product of the y values for the points $x = 2$ and $x = 3$ is less than zero. In function notation, this requirement is written $f(2) * f(3) < 0$. For the other point we also have $f(5) * f(6) < 0$. In both cases the product of two factors of opposite sign is negative. This happens only when roots are between the values. For other values, like $f(1) * f(2)$ and $f(4) * f(5)$, the products are positive.[2] This is the key fact that will support our binary search.

Here is a program (not a very good one, but we will improve it later) that finds the root of our equation. To use it, you enter the two x values between which you wish to find a root. For example, you can let $P = 2$ and $Q = 3$ to find one root and then run the program again letting $P = 5$ and $Q = 6$ to find the other.

```
PROGRAM:BISCH1
: Prompt P,Q
: P→X
: X²−8X+13→Y
: Y→L
: Q→X
: X²−8X+13→Y
```

[2]Alert readers will notice that another situation can occur. It is possible that there is a root at a point where the function is tangent to the x axis. To simplify this discussion, I avoid this complication until the end of this appendix.

```
: Y→R
: While Q ≠ P
:     (P+Q)/2→M
:     M→X
:     X²−8X+13→Y
:     Y→N
:     If L*N<0        (I)
:         Then
:                 M→Q
:             Else
:                 M→P
:     End (If)
: End (While)
: Disp M
```

The line marked (I) is the key to this program. It represents the test to see where the graph crosses the x axis.

Here is how the program works. Its first part determines the corresponding y values for the entered x values. In effect, we determine the points (P, L) and (Q, R). We know that the graph crosses the x axis between those points, so we know that $L * R < 0$.

Now we enter the While loop. In it we determine a new x value, half way between P and Q. We plug it into our function to determine its y value. This way we create the new point (M, N). With the If test, we check the y value of this point against the y value of (P, L). If their product is negative, we know that the root is between P and M, so we assign to Q this M value. If not, we know the root must lie between M and Q. In this case we assign M to P.

In our example, we would have calculated $(P, L) = (2, 1)$ and $(Q, R) = (3, -2)$ before we first enter the While loop. In that loop the midpoint M is 2.5 and we calculate the corresponding y value to give $(M, N) = (2.5, -.75)$. Since L is positive and N is negative, we know that the root is between P and M. For that reason we assign M to Q and proceed. At that point we have the root between P and Q again, but P and Q are only half as far apart.

If you run this program with $P = 2$ and $Q = 3$, you will find the root $x = 2.267949192$. If you run it with $P = 5$ and $Q = 6$, you will find the other root, $x = 5.732050808$. The quadratic formula tells us that the roots of this equation are

$$x = \frac{8 \pm \sqrt{(-8)^2 - 4 * 1 * 13}}{2 * 1} = 4 \pm \sqrt{3}$$

These give us the same values for x.

So we have solved the problem, but suppose now that we want to use the program to solve another equation. To do so, we would have to type the new function into three separate lines (lines 3, 6, and 11). That's not at all efficient, and we will show in the following section how this can be avoided.

FUNCTIONS, SUBROUTINES AND PROGRAM CALLING

To accomplish our purpose, we will need to use one of several possible programming tricks, not all of which are available on basic programmable calculators. We'll take them in order of sophistication.

First, let's reinforce what the problem is. We have seen that our binary search program works. All we want to do now is find a way to use it to solve any equation simply by typing that new equation once.

The simplest way to do this would be to define a function, say, $FN(X) = X^2 - 8X + 13$. Then the program could read as follows:[3]

```
PROGRAM:BISCH2
: Prompt P,Q
: P→X
: FN(X)→Y
: Y→L
: Q→X
: FN(X)→Y
: Y→R
: While Q ≠ P
:     (P+Q)/2→M
:     M→X
:     FN(X)→Y
:     Y→N
:     If L*N<0
:           Then
:                   M→Q
:            Else
:                   M→P
:    End (If)
: End (While)
: Disp M
```

Then, if you want to solve another equation, simply redefine the function $FN(X)$ and again run the program. Unfortunately, this method is not available in many elementary calculators, including the TI-84.

---

[3]Depending on the calculator or computer, the definition could be included either in the program or separate from it.

A second approach to this problem is through what is called a *subroutine*. Here is how the program might appear:

```
PROGRAM:BISCH3
: Prompt P,Q
: P→X
: Gosub FN
: Y→L
: Q→X
: Gosub FN
: Y→R
: While Q ≠ P
:    (P+Q)/2→M
:    M→X
:    Gosub FN
:    Y→N
:    If L*N<0
:          Then
:                M→Q
:          Else
:                M→P
:    End (If)
: End (While)
: Disp M
:
: Subroutine FN
: X²−8X+13→Y
: Return
```

What is happening in this program is that the key line evaluating the given function is included just once in a separate program section called a *subroutine*, in this case `Subroutine FN`. Whenever program operation reaches a line instruction that says `Gosub FN`, control is sent to that subroutine and the instruction is carried out. The `Return` command then returns control to the next step after the `Gosub` instruction. Once again, unfortunately, this structure is not available in elementary calculators, including the TI-84.

There is third means, however, that is available. It involves establishing another program that is just like the subroutine. Because this new program is so much like the others, we will call it `FN`. Here is how its two steps would appear:

```
PROGRAM:FN
: X²−8X+13→Y
: Return
```

Now we revise the original program to read

```
PROGRAM:BISCH4
: Prompt P,Q
: P→X
: prgmFN⁴
: Y→L
: Q→X
: prgmFN
: Y→R
: While Q ≠ P
:    (P+Q)/2→M
:    M→X
:    prgmFN
:    Y→N
:    If L*N<0
:        Then
:                M→Q
:            Else
:                M→P
:    End (If)
: End (While)
: Disp M
```

If you have another equation you wish to solve, you need only change that single instruction in the program FN, then run BISCH4.

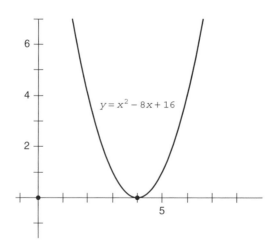

Figure M.2   The graph of $y = x^2 - 8x + 16$.

⁴The instruction prgm is found in the TI-84's "CATALOG".

Depending on the calculator or computer available to you, you can use one of these methods to provide a binary search program to locate roots of complex equations. Many calculators include built-in programs that solve such equations, but now you have some insight into how that can be done.

I conclude this appendix by underscoring the reservation mentioned earlier in footnote 2. There are equations with roots that cannot be found by the method outlined here. A simple example is $x^2 - 8x + 16 = 0$ (see Figure M.2). Since this is equivalent to the equation $(x - 4)^2 = 0$, inspection tells us that a root is 4.

The problem is that $y = x^2 - 8x + 16$ does not cross the x axis, so we cannot use the sign change test of the binary search technique. This graph (and others) have what are called *multiple roots*, where they are tangent to the x axis. Some other technique is necessary to zero in on such roots.

APPENDIX N

DERIVATION OF THE LOGARITHM CHANGE OF BASE FORMULA

We set out to prove the logarithm change of base formula:

$$\log_b x = \frac{\log_a x}{\log_a b}$$

To do so, we let $y = \log_b x$ and apply these as exponents on the base $b : b^y = b^{\log_b x}$

By log property (I) of page 87, the right side of this equation is simply x. Thus we have $b^y = x$.

We take $\log_a$ of each side of this equation, which gives us $\log_a b^y = \log_a x$, and then we apply the log rule for exponents to the left side to produce $y \log_a b = \log_a x$.

Now we substitute the value of y that we established at the outset. This gives us

$$\log_b x * \log_a b = \log_a x$$

Dividing each side of this equation by $\log_a b$, we have the desired formula

$$\log_b x = \frac{\log_a x}{\log_a b}$$

Inside Your Calculator: From Simple Programs to Significant Insights   By Gerald R. Rising
Copyright © 2007 John Wiley & Sons, Inc.

APPENDIX O

THE RATIO OF DECIMAL TO BINARY DIGITS

One of the tradeoffs in changing from decimal to binary representation is an increase in the number of digits. For example, we have $93_{ten} = 1011101_{two}$. In this case we trade two digits for seven; thus we have $3\frac{1}{2}$ times as many digits. If we examine additional values, we can construct the following table:

Binary	Decimal	Binary–Decimal	* Increase
1	1	1–1	1
10	2	2–1	2
100	4	3–1	3
1000	8	4–1	4
10000	16	5–2	2.5
100000	32	6–2	3
1000000	64	7–2	3.5
10000000	128	8–3	~3.7
100000000	256	9–3	3
1000000000	512	10–3	~3.3
10000000000	1024	11–4	2.75

Although these values vary a good deal, it may be that they approach a limit as they continue to increase. In exploring this problem, we will use

Inside Your Calculator: From Simple Programs to Significant Insights   By Gerald R. Rising
Copyright © 2007 John Wiley & Sons, Inc.

logarithms with bases 2, e and 10. In the discussion that follows, the usual abbreviations will apply: $\log_2 = \lg$, $\log_e = \ln$, and $\log_{10} = \log$.

Consider an integer D. Experimentation will show that the number of decimal digits in D is $1 + \text{int}(\log N)$. For example, the number of digits in 8500 is $1 + \text{int}(\log 8500) = 1 + \text{int}(3.929418926) = 1 + 3 = 4$.

Also, the number of digits in the binary representation of D is $1 + \text{int}(\lg D)$. For our decimal number 8500, this would give us $1 + \text{int}(\lg 8500)$. We can calculate $\lg 8500$ by the change of base formula as

$$\lg 8500 = \frac{\ln 8500}{\ln 2} = \frac{9.047821442}{.6931471806} = 13.05324713$$

and substitute this in the previous line to give $= 1 + \text{int}(13.05324713) = 1 + 13 = 14$.

We can check this result by converting 8500 to binary and simply counting the digits: $8500_{\text{ten}} = 10000100110100_{\text{two}}$.

It is worth pointing out here that $\text{int}(\log N)$ and $\text{int}(\lg N)$ are in each case the same as the characteristic of the logarithm. In any numeration system the number of digits is one more than the characteristic of its logarithm.[1]

In any case the limit we seek is[2]

$$\lim_{N \to \infty} \frac{1 + \text{int}(\lg N)}{1 + \text{int}(\log N)}$$

in which we can substitute the change of base formulas using $\ln$ to give:[3]

$$\lim_{N \to \infty} \frac{1 + \text{int}(\lg N)}{1 + \text{int}(\log N)} = \lim_{N \to \infty} \frac{1 + \text{int}\left(\dfrac{\ln N}{\ln 2}\right)}{1 + \text{int}\left(\dfrac{\ln N}{\ln 10}\right)}$$

In this expression, as N gets large, the 1s become insignificant. So, too, do the mantissas of the logarithms, because the characteristics get larger and

---

[1] I have extended the use of the word "characteristic" from base 10 logs to include the integer part of any log.

[2] Those of you who are unfamiliar with the notation $\lim_{N \to \infty}$ can think of it informally as indicating a number approached when N gets very large.

[3] Alert readers will notice that we could continue here using $\log$ instead of $\ln$. It is an interesting exercise to finish the argument in this way. The result will be $(\log 10)/\log 2$, which is equal to $(\ln 10)/\ln 2$.

larger while the mantissas stay between zero and one. For these reasons we have a much simpler expression

$$\lim_{N \to \infty} \frac{\dfrac{\ln N}{\ln 2}}{\dfrac{\ln N}{\ln 10}} = \lim_{N \to \infty} \frac{\ln 10}{\ln 2}$$

and, since this last fraction does not involve N, it is equivalent to

$$\frac{\ln 10}{\ln 2} \approx 3.32$$

What this tells us is that, while the number of binary digits representing numbers is greater than the number of decimal digits, even when the base 10 numbers get very large, their binary equivalents get closer and closer to about $3\frac{1}{3}$ times as large. In fact, the most they ever get was back when we had single digits. Then for two specific values the number of binary digits is four times as great: $8_{\text{ten}} = 1000_{\text{two}}$ and $9_{\text{ten}} = 1001_{\text{two}}$.

APPENDIX P

CONSTRUCTING A LOG TABLE

> The satisfaction of owning an electronic calculator and the fun of
> experimenting with its buttons hardly measure up to the pride felt
> possessing and mastering an advanced slide rule.
>
> —Jan Gullberg

Shortly after the invention of logarithms, two English mathematicians
contributed to the construction of the first slide rules. In 1620 Edmund
Gunter developed a logarithmic scale and used compasses to multiply and
divide with it. A year later William Oughtred paired two of these scales
to construct the first slide rule.

It is an interesting exercise to develop a table of logarithms from which
a pair of scales can then be constructed to provide a simple slide rule.

We begin with the equally spaced decimal ruler of Figure P.1.

We will enter a log scale above these units, and, as a first step, we enter
the numbers whose logs are 0 and 1 above those numbers on the scale to
produce Figure P.2.

That was rather straightforward, but now we want to locate 2, 3, 4,
5, 6, 7, 8, and 9 on that scale. We could do this by using log tables,

Inside Your Calculator: From Simple Programs to Significant Insights   By Gerald R. Rising
Copyright © 2007 John Wiley & Sons, Inc.

Figure P.1   A decimal ruler.

Figure P.2   Beginning the log scale.

but it is instructive to see how we can do this simply by using the basic principles of logs. (This is one of those problems that you could address when marooned on a desert island.)

The one other thing we have to work with is our knowledge of the logs of powers of 10: $\log 1 = 0, \log 10 = 1, \log 100 = 2, \log 1000 = 3$, and so on.

SEVERAL LOGARITHM VALUES

Log 2

What we can do is seek a power of 2 near one of those powers of 10. We know $2^3 = 8$, which is not very near 10; and $2^6 = 64$ and $2^7 = 128$, neither of which is very close to 100. But $2^{10} = 1024$ is quite close to 1000. We can then write this relationship as $2^{10} = 1000 * 1.024$, and we can write the log equation for this: $10 \log 2 = \log 1000 + \log 1.024$.

We know $\log 1000 = 3$, so we have $10 \log 2 = 3 + \log 1.024$, and dividing by 10 gives us

$$\log 2 = .30 + \frac{\log 1.024}{10}$$

This tells us that $\log 2 = .30$ plus an error term. We know that $\log 1 = 0$, so $\log 1.024$ will be small, and here even that is divided by 10. In fact, for any positive x, since $1 + x < 10^x, \log(1 + x) < x$. In this case that means that $\log 1.024 < .024$. When we divide that by 10, our error is less than .0024.

Thus we have $\log 2 \approx .30$ and we can place this on our scale as shown in Figure P.3.

Figure P.3   Placing 2 on the scale.

Figure P.4   Placing 4, 5 and 8 on the scale.

Log 4, Log 5, and Log 8

Knowing the location of log 2, we can also locate log 4 and log 8:

$$\log 4 = \log 2^2 = 2 \log 2 \approx 2 * .30 \approx .60$$

$$\log 8 = \log 2^3 = 3 \log 2 \approx 3 * .30 \approx .90$$

And

$$\log 5 = \log(10/2) = \log 10 - \log 2 \approx 1 - .30 = .70$$

so we have Figure P.4.

Log 3, Log 6, and Log 9

In estimating log 3, we have the values that we have already calculated in addition to logs of powers of 10. We can write $3^4 = 81 = 80 * 81/80 = 8 * 10 * 1.0125$, and taking logs of the first and last of these, we have $4 * \log 3 = \log 8 + \log 10 + \log 1.0125$ or $4 * \log 3 \approx .9 + 1 + \log 1.0125$, and, dividing by 4:

$$\log 3 \approx .475 + \frac{\log 1.0125}{4}$$

So we have $\log 3 \approx .48$ and this time also our error term:

$$\frac{\log 1.0125}{4} < \frac{.0125}{4} \approx .003.$$

From $\log 3 \approx .48$ we get other values:

$$\log 6 = \log 2 + \log 3 \approx .3 + .48 \approx .78$$

$$\log 9 = \log 3^2 = 2 * \log 3 \approx 2 * .48 = .96$$

Placing these on the scale, we have Figure P.5.

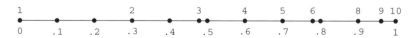

Figure P.5   Adding 3, 6, and 9 to the scale.

Figure P.6   Adding 7 to the scale.

Figure P.7   A log scale.

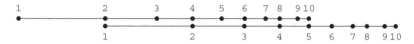

Figure P.8   Using log scales for multiplication by 2.

Log 7

Finally, we begin with $7^2 = 49 = 50 * 49/50 = 5 * 10 * 49/50$. Taking the log equation for the first and last members we have $2 * \log 7 = \log 5 + \log 10 + \log .98 = .70 + 1 + \log .98 = 1.7 + \log .98$ and, dividing by 2, we have

$$\log 7 = .85 + \frac{(\log .98)}{2}$$

Thus we have $\log 7 \approx .85$, with an error of $(\log .98)/2$. This error estimate is $.02/2 = .01$, so we have an error of less than 1%.

Placing this on our scale, we have Figure P.6.

Now we can take away the values below the scale line and we have the log scale of Figure P.7.

If we make a copy of this scale, we can use it to perform simple multiplications. Figure P.8 shows multiplication by 2 by shifting the second scale so that it begins at 2 on the first scale. (The scales are shortened proportionally to allow this.)

Note that the multiples of 2 are read above the corresponding values on the lower scale.

You can also multiply by 5 with the scale placed like this. Since the 5 is placed below the 10 (usually recorded as 1 on slide rules), you can read products of 5 times numbers on the top scale below them. Of course, $5 * 6$ is 30, not 3, but one of the tasks of the slide rule user is to place the decimal point correctly. All this is because the slide rule is proportional to log mantissas, and the characteristics must be taken into account by the user.

Clearly, then, the rules as placed in Figure P.8 also provide products like $20 * 300$ and $.05 * 70$.

There are several interesting avenues to explore from here. It should be evident that additional values can be added to the scales from their current placement. For example, as the scales are placed in Figure P.8, we can locate 2.5, 3.5, and 4.5 on the lower scale below their doubles on the upper scale.

So far we have constructed a log table for the numbers from 1 to 10:

X	Log X
1	.00
2	.30
3	.48
4	.60
5	.70
6	.78
7	.85
8	.90
9	.96
10	1.00

If you look at the scales already constructed, you should notice that few values are filled in at the left end of the scales. You can continue to calculate values from 11 through 19 (which correspond to $1.1-1.9$) to fill in many points there.

To do this, you can use:

$$
\begin{aligned}
11^2 &= 121 = 120 * 121/120 = 3 * 4 * 10 * 121/120 \\
12 &= 3 * 4 \\
13 &= 40/3 * 39/40 \\
14 &= 2 * 7 \\
15 &= 3 * 5 \\
16 &= 2^4 \\
17^3 &= 4913 = 7^2 * 100 * 4913/4900 \\
18 &= 3 * 6 \\
19^2 &= 361 = 6^2 * 10 * 361/360
\end{aligned}
$$

APPENDIX Q

RELATIONS BETWEEN SIDES OF INSCRIBED AND CIRCUMSCRIBED POLYGONS

In this appendix we set out to establish those two formulas that relate the lengths of sides of inscribed and circumscribed regular polygons as they are doubled (Figure Q.1).

The formula for Figure Q.1 that we seek to establish for this process is $S_{2n} = \sqrt{2 - \sqrt{4 - S_n{}^2}}$.

The corresponding formula for doubling the number of sides of circumscribed polygons as on Figure Q.2 is

$$T_{2n} = \frac{2\sqrt{T_n^2 + 4} - 4}{T_n}$$

To accomplish these tasks, we must first review some relationships from elementary geometry:[1]

1. In a circle two inscribed angles are equal if they intercept equal arcs. In Figure Q.3, if arc a = arc b, then angle A = angle B and vice versa.

2. As we have pointed out before, an angle inscribed in a semicircle is a right angle. In Figure Q.4, angle A is a right angle.

---

[1]This development follows Archimedes' work according to Sherman Stein's account in *Archimedes: What Did He Do Besides Cry Eureka?* (Mathematical Association of America, 1999).

Inside Your Calculator: From Simple Programs to Significant Insights   By Gerald R. Rising
Copyright © 2007 John Wiley & Sons, Inc.

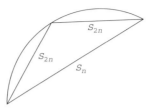

Figure Q.1   Doubling the number of sides of inscribed regular polygons.

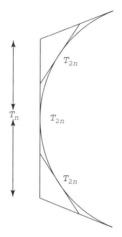

Figure Q.2   Doubling the number of sides of circumscribed regular polygons.

3. An angle bisector of a triangle divides the opposite side in the same ratio as the corresponding sides of the angle. This theorem tells us that on Figure Q.5, with angle BAC bisected, $(BD/DC) = (AB/AC)$.

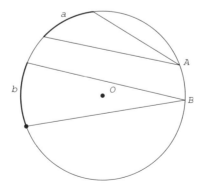

Figure Q.3   Inscribed angles and their arcs.

Because this theorem is often omitted from geometry instruction today, an outline of its proof is shown in Figure Q.6.

Construct a line parallel to DA through C and extend BA to meet this line at E. Then parallel line relationships make all the angles marked x equal. Thus $AC = AE$. From similar triangles ADB and ECB, we have $(BD/DC) = (BA/AE)$ and, substituting AC for AE, this becomes $(BD/DC) = (BA/AC)$, as we set out to prove.

4. In proportions, if $a/b = c/d$, then

$$\frac{a}{b} = \frac{a+c}{b+d}$$

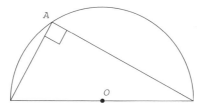

Figure Q.4   A right angle inscribed in a semicircle.

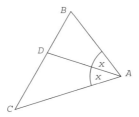

Figure Q.5   Triangle angle bisector theorem.

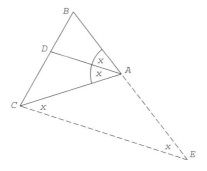

Figure Q.6   Triangle angle bisector theorem proof.

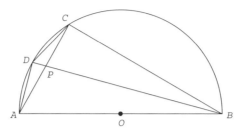

Figure Q.7 Doubling sides.

Since $(a/b) = (c/d)$, $ad = bc$. Adding ab to each member, we have $ab + ad = ab + bc$, which may be written $a(b + d) = b(a + c)$. This last is equivalent to the desired

$$\frac{a}{b} = \frac{a + c}{b + d}$$

Now we are prepared to prove those two formulas. First we address the inscribed regular polygons.

In the semicircle of Figure Q.7, we have $AC = S_n$ and $AD = DC = S_{2n}$. Because they are inscribed in the semicircle, angles ADB and ACB are both right angles and inscribed angles ABD and DBC are equal because they intercept equal arcs. Therefore, right triangles ADB and PCB are similar, and we have proportional sides:

$$\frac{BD}{AD} = \frac{BC}{PC} \tag{I}$$

Since BP bisects angle B in triangle ABC, the angle bisector theorem we just proved gives us another proportion: $AB/BC = AP/PC$ or, exchanging means:

$$\frac{AB}{AP} = \frac{BC}{PC} \tag{II}$$

Now we apply that relationship (4) about proportions on page 267 to (II) to give

$$\frac{AB}{AP} = \frac{AB + BC}{AP + PC}$$

and this is equivalent to

$$\frac{AB}{AP} = \frac{AB + BC}{AC} \tag{III}$$

Since the right sides of the proportions marked (I) and (II) are the same, their left members are equal. Thus $BD/AD = AB/AP$, and we can substitute this into (III) to obtain

$$\frac{BD}{AD} = \frac{AB + BC}{AC} \tag{IV}$$

Archimedes can calculate all the terms on the right side of that equation, because $AC = S_n$, $AB = 2$, and he can calculate BC by the Pythagorean theorem. For that reason, at this point he knows the ratio BD/AD.

Again the Pythagorean theorem gives him $AB = \sqrt{AD^2 + BD^2}$ and, dividing by AD, he has

$$\frac{AB}{AD} = \frac{\sqrt{AD^2 + BD^2}}{AD}$$

which may be rewritten as

$$\frac{AB}{AD} = \sqrt{1 + \left(\frac{BD}{AD}\right)^2} \tag{V}$$

He can insert that ratio BD/AD into this relationship to obtain AB/AD, and knowing $AB = 2$, he can finally calculate $AD = S_{2n}$.

You can just imagine his having to work out all those ratios without having algebraic language to help him, and he had to do all those calculations over each time he doubled the side lengths! No wonder he quit after doing so four times.

Here, then, is the algebra that backs up those calculations and allows us to use a calculator to carry them out.

We begin with $AB = 2$ and $AC = S_n$. From this we calculate, as above, $BC = \sqrt{4 - S_n^2}$.

Now, we can evaluate equation (IV):

$$\frac{BD}{AD} = \frac{AB + BC}{AC} = \frac{2 + \sqrt{4 - S_n^2}}{S_n}$$

and substitute this ratio into (V) to give us

$$\frac{AB}{AD} = \sqrt{1 + \left(\frac{BD}{AD}\right)^2} = \sqrt{1 + \left(\frac{2 + \sqrt{4 - S_n^2}}{S_n}\right)^2}$$

Squaring and combining terms produces a somewhat simpler radical

$$\frac{AB}{AD} = \frac{2\sqrt{2 + \sqrt{4 - S_n^2}}}{S_n}$$

Now recall from our diagram that $AB/AD = 2/S_{2n}$ and we have

$$\frac{2}{S_{2n}} = \frac{2\sqrt{2 + \sqrt{4 - S_n{}^2}}}{S_n}$$

Inverting, multiplying by 2, and finally rationalizing that denominator by multiplying numerator and denominator by $\sqrt{2 - \sqrt{4 - S_n{}^2}}$ provides the first of our formulas relating S_n to S_{2n} for inscribed regular polygons: $S_{2n} = \sqrt{2 - \sqrt{4 - S_n{}^2}}$.

Although Archimedes used essentially this approach in his estimate of π, there are problems with this formula, which will be discussed in Appendix R. In that appendix a substitute for this formula will be determined that responds to those problems:

$$S_{2n} = \frac{S_n}{\sqrt{1 + \frac{S_n}{2}} + \sqrt{1 - \frac{S_n}{2}}}$$

For now, however, we turn to the second problem of this appendix, to derive the equation for doubling the number of sides of circumscribed regular polygons:

$$T_{2n} = \frac{2\sqrt{T_n^2 + 4} - 4}{T_n}$$

Since the 6 sides of the circumscribed regular hexagon of the left side of Figure Q.8 are intercepted by $60°$ central angles, $AB = T_6$ and, since the twelve sides of the regular dodecagon are intercepted by $30°$ angles,

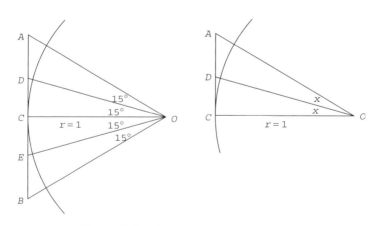

Figure Q.8 Circumscribed polygon angles.

$DE = T_{12}$. The right figure generalizes this for half-sides of the circumscribed polygons. With the angle at O bisected, we let $AC = T_n/2$ and $DC = T_{2n}/2$.

On this right-hand figure, at each stage of the doubling process, Archimedes knows all three sides of triangle OAC. He knows $OC = 1$, the radius of the unit circle; $AC = T_n/2$; and he can calculate OA by the Pythagorean theorem:

$$OA = \sqrt{\left(\frac{T_n}{2}\right)^2 + 1}$$

Once again we make use of that angle bisector theorem to get us started. On triangle OAC, we have

$$\frac{OA}{OC} = \frac{AD}{CD}$$

Now we simply manipulate proportions, first adding one (in the form of OC/OC and CD/CD) to each member. This is equivalent to adding 1 to each side of the equation, and thus equality is maintained:

$$\frac{OC}{OC} + \frac{OA}{OC} = \frac{CD}{CD} + \frac{AD}{CD}$$

Then, combining fractions, we obtain

$$\frac{OC + OA}{OC} = \frac{CD + AD}{CD}$$

and finally substituting CA for $CD + AD$, we arrive at

$$\frac{OC + OA}{OC} = \frac{CA}{CD}$$

Since Archimedes knows all the other values in this proportion, he can now calculate CD.

We can replicate this algebraically by substituting the values we already know in this last expression to obtain

$$\frac{\sqrt{\left(\frac{T_n}{2}\right)^2 + 1} + 1}{1} = \frac{\frac{T_n}{2}}{\frac{T_{2n}}{2}}$$

This we solve for T_{2n} and simplification (including rationalizing the denominator) leads us to the desired formula:

$$T_{2n} = \frac{2\sqrt{T_n^2 + 4} - 4}{T_n}$$

APPENDIX R

CHANGE IN FORM OF A POLYGON FORMULA

We showed in Appendix Q the derivation of the formula associated with doubling the number of sides of inscribed regular polygons. That formula is $S_{2n} = \sqrt{2 - \sqrt{4 - S_n{}^2}}$.

There are problems with this formula: (1) it has a square root within a square root; and (2) when S_n becomes small, calculating problems arise. When you subtract a very small value from 4, you begin to lose accuracy.

To see how this happens, consider a trivial example. Suppose $S_n =$ .0000001111111111. (Your calculator would carry all of those ten digits in the form 1111111111E$-$7). But now when you subtract that number from 4, and you get (again to ten-digit accuracy), 3.999999889, which is the same as $4 - .000000111$. The rest of those digits were lost in the processing. (Even though your calculator probably carries a few extra digits internally, this effect still holds but just at another level.)

We will address these problems by converting that formula to another form:

$$\sqrt{2 - \sqrt{4 - S_n{}^2}} = \frac{S_n}{\sqrt{1 + \frac{S_n}{2}} + \sqrt{1 - \frac{S_n}{2}}}$$

We show this in two steps. In the first step we use a little-known theorem that, for a particular situation, "denests" a square root that occurs as a

Inside Your Calculator: From Simple Programs to Significant Insights   By Gerald R. Rising
Copyright © 2007 John Wiley & Sons, Inc.

term within another square root. It states: When $a^2 - b = p^2$ for some $p > 0$, then

$$\sqrt{a \pm \sqrt{b}} = \sqrt{\frac{a+p}{2}} \pm \sqrt{\frac{a-p}{2}}$$

This theorem is proved by squaring that equation, making appropriate substitutions, and simplifying the result.

We apply the theorem to our situation with $a = 2$ and $b = 4 - S_n^2$, since $a^2 - b = 4 - (4 - S_n^2) = S_n^2$. We then set $S_n^2 = p^2$ to get $p = S_n$. Since in this case the $\pm$ of the theorem is negative, we arrive at the following equation:

$$\sqrt{2 - \sqrt{4 - S_n^2}} = \sqrt{1 + \frac{S_n}{2}} - \sqrt{1 - \frac{S_n}{2}}$$

There is still more to do, however, since we are still subtracting two nearly equal quantities when S_n is small.

In the second step, we rationalize the numerator of the right side of that equation by multiplying by

$$\frac{\sqrt{1 + \frac{S_n}{2}} + \sqrt{1 - \frac{S_n}{2}}}{\sqrt{1 + \frac{S_n}{2}} + \sqrt{1 - \frac{S_n}{2}}}$$

Here is the math:

$$\sqrt{1 + \frac{S_n}{2}} - \sqrt{1 - \frac{S_n}{2}} * \frac{\sqrt{1 + \frac{S_n}{2}} + \sqrt{1 - \frac{S_n}{2}}}{\sqrt{1 + \frac{S_n}{2}} + \sqrt{1 - \frac{S_n}{2}}} = \frac{\left(1 + \frac{S_n}{2}\right) - \left(1 - \frac{S_n}{2}\right)}{\sqrt{1 + \frac{S_n}{2}} + \sqrt{1 - \frac{S_n}{2}}}$$

$$= \frac{S_n}{\sqrt{1 + \frac{S_n}{2}} + \sqrt{1 - \frac{S_n}{2}}}$$

These successive steps have shown that

$$\sqrt{2 - \sqrt{4 - S_n^2}} = \frac{S_n}{\sqrt{1 + \frac{S_n}{2}} + \sqrt{1 - \frac{S_n}{2}}}$$

Substitution then gives us the alternate formula for S_{2n} that we sought:

$$S_{2n} = \frac{S_n}{\sqrt{1 + \frac{S_n}{2}} + \sqrt{1 - \frac{S_n}{2}}}$$

Notice what happens in this formula when S_n gets very small. At that time the denominator gets closer and closer to $\sqrt{1} + \sqrt{1} = 2$. Thus $S_{2n} \approx S_n/2$. However, we have twice as many sides in the new polygon and $P_{2n} \approx 2 * P_n/2 \approx P_n$, the situation you should expect to be true when the number of sides is large. As the polygon perimeters approach the circumference of the circle, they change less and less each time the number of sides is doubled.

APPENDIX S

AN AREA APPROACH
TO ARCHIMEDES' PROBLEM

My colleague Donald Stover has brought to my attention an entirely different approach to Archimedes' problem of evaluating π. Instead of working with the circumferences of inscribed and circumscribed polygons, Stover uses the reciprocals of their areas in the following program:

```
PROGRAM:STOVEPI
: .5→I
: .25→C
: While I ≠ C
:     √(I*C)→I
:     (I+C)/2→C
: End (While)
: Disp 1/I
```

This program is, I suggest, quite remarkable. It not only is brief but it also involves two old friends, the geometric mean in line 4 and the arithmetic mean in line 5. I consider the basis for this program the most remarkable algorithm of this book.

It may be a short and seemingly straightforward program, but working it out is not a trivial task. I will provide only an outline of that math here.[1]

---

[1]For a complete development, I refer you to Chapter 5 of Stover's book, *Precalculus Problems and Projects*, which is in the final stages of preparation.

Inside Your Calculator: From Simple Programs to Significant Insights   By Gerald R. Rising
Copyright © 2007 John Wiley & Sons, Inc.

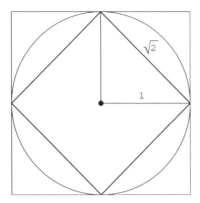

Figure S.1 Inscribed and circumscribed squares.

It is important to understand that in this program the variable I (for inscribed) represents the *reciprocal* of the area of a polygon inscribed in a unit circle and C (for circumscribed) represents the *reciprocal* of the area of the corresponding circumscribed polygon.

The program begins with inscribed and circumscribed squares as shown in Figure S.1. It should be evident from this diagram that the area of the inscribed square is 2 and the area of the circumscribed square is 4. Thus the program begins with their reciprocals, $I = 1/2$ and $C = 1/4$.

In the `While` loop these area reciprocals are recalculated as the number of sides of the polygons are doubled just as the perimeters were in Archimedes' approach.

Here are the corresponding formulas in which n represents the number of sides of the polygon and x represents half the length of the side of the inscribed polygon, in both cases before doubling the number of sides.

AREA FORMULAS

	Inscribed Polygon	Circumscribed Polygon
n-sided	$nx\sqrt{1-x^2}$	$\dfrac{nx}{\sqrt{1-x^2}}$
$2n$-sided	nx	$\dfrac{2nx}{1+\sqrt{1-x^2}}$

Deriving the formulas for the n-sided polygon is a reasonable exercise, but doing so for the $2n$-sided polygon represents a challenging problem. Once you have those formulas, however, forming their reciprocals to give the following table is straightforward.

AREA FORMULA RECIPROCALS

	Inscribed Polygon	Circumscribed Polygon
n-sided	$\dfrac{1}{nx\sqrt{1-x^2}} = \text{old } I$	$\dfrac{\sqrt{1-x^2}}{nx} = \text{old C}$
$2n$-sided	$\dfrac{1}{nx} = \text{new } I$	$\dfrac{1+\sqrt{1-x^2}}{2nx} = \text{new C}$

It is also straightforward to show that

$$\sqrt{\text{old } I * \text{old } C} = \text{new } I \quad \text{and} \quad \frac{\text{new } I + \text{old } C}{2} = \text{new C}$$

Those are the formulas that operate in each pass through the While loop. It is important to notice that the arithmetic mean is calculated with the new I. Thus you don't have to create a dummy variable to retain the old I for this calculation.

To show the convergence of values, you can modify the program to this form:

```
PROGRAM STOVPI2
: .5→I
: .25→C
: 4→N
: Disp N,2,4
: Pause
: While I ≠ C
:    √(I*C)→ I
:    (I+C)/2→C
:    Disp N,1/I,1/P
:    Pause
: End (While)
```

This algorithm may be used with computer algebra systems that allow high-precision arithmetic to calculate π to many more digits. It does not address the problem that our program that mirrored Archimedes' method did and when run on your calculator converges on the calculator's 10-digit value of π. In the following scheme accurate digits are underscored.

I	N	$1/I$	$1/C$	Correct Digits
1	4	2	4	0
2	8	2.828427125	**3**.313708499	0
3	16	3.061467459	3.182597878	1
4	32	3.121445152	3.151724907	1
5	64	3.136548491	3.144118385	3
6	128	3.140331157	3.142223630	3
7	256	3.141277251	3.141750369	3
8	512	3.141513801	3.141632081	4
9	1024	3.141572940	3.141602510	5
10	2048	3.141587725	3.141595118	5
11	4096	3.141591422	3.141593270	6
12	8192	3.141592346	3.141592808	6
13	16384	3.141592577	3.141592692	7
14	32768	3.141592634	3.141592663	7
15	65536	3.141592649	3.141592656	7
16	131072	3.141592652	3.141592654	9
17	262144	3.141592653	3.141592654	9
18	524288	3.141592654	3.141592654	10

FURTHER READING:
A PERSONAL SELECTION

Benson, D. C. (1999). *The Moment of Proof: Mathematical Epiphanies*. New York: Oxford Universiy Press. (A clear development of how mathematical ideas are proved through many interesting and carefully explained examples. A wonderful book that requires only school algebra. Sets a high standard for science exposition.)

Berlinski, D. (2001). *The Advent of the Algorithm: The Idea that Rules the World*. New York: Harcourt, Inc. (The development of logic over three centuries leads to today's computing languages and programs. A bit cute but generally entertaining and a useful history.)

Dantzig, T. (1930). *Number: The Language of Science*. New York: The Free Press. (Long considered a classic. This book belongs in the library of everyone interested in mathematics.)

Dijksterhuis, E. J. (1987). *Archimedes*. Princeton, NJ, Princeton University Press. (A serious and demanding survey of Archimedes' ideas.)

Dozenal Society of America. "The Dozenal Society," available from `www.polar.synynassau.edu/~dozenal/contact.html`. (If only our hands had six fingers! There is a strong case for use of base 12, and it is reasonable to consider it via this website. We are, however, too deeply invested in base 10 to change.)

Dunham, W. (1990). *Journey Through Genius: The Great Theorems of Mathematics*. New York: Wiley Science Editions. (Key theorems of mathematicians from Hippocrates and Pythagoras to Euler and Cantor well presented and placed in historical context. An excellent introduction to the nature of proof.)

Engel, A. (1984). *Elementary Mathematics from an Algorithmic Standpoint*. Staffordshire, UK: Keele Mathematical Education Publications, University of

Keele. (A remarkable German teacher gathered many algorithms in this book written early in the time of calculators and computers. Anything this outstanding expositor developed is worth study. Some of the algorithms of *Inside Your Calculator* are modified from Engel's ideas.)

Engel, A. (1993). *Exploring Mathematics with Your Computer*. Washington, DC, Mathematical Association of America. (Another original presentation by this brilliant teacher. Unfortunately, the programs are written in Pascal, a language no longer in favor. Interpreting the programs is worth the effort, however. Extends the ideas of *Inside Your Calculator*.)

Eves, H. (1964). *An Introduction to the History of Mathematics*. New York: Holt, Rinehart and Winston. (Although this book is old, I know of no math history that involves the reader as this one does. The exercise sets take you close to the problems solved by historically important mathematicians.)

Flannery, D. (2006). *The Square Root of 2: A Dialogue Concerning a Number and a Sequence*. New York: Copernicus Books. (A master teacher takes a serious sudent through a series of discoveries related to root 2. An interesting and comfortable way to learn a great deal about this interesting number.)

Gullberg, J. (1997). *Mathematics from the Birth of Numbers*. New York: Norton & Company. (If you have the time to study just one math book, choose this one. It is an encyclopedic survey of math well presented in historical order. Heavier going toward the end (of over a thousand pages), but the learning curve is reasonably inclined.)

Heath, T. L., Ed. (1897). *The Works of Archimedes*. London, Cambridge University Press. (An early survey of Archimedes' writing by a world authority on the mathematics of antiquity.)

Heath, T. L. (1963). *A Manual of Greek Mathematics*. New York, Dover Publications, Inc. (A world authority's survey of the amazing work ancient mathematicians accomplished with weak tools and rotten notation. Much is remarkably applicable today.)

Hofstadter, D. R. (1999). *Godel Escher Bach: An Eternal Golden Braid*, New York: Basic Books, Inc. (The 20th anniversary edition of a Pulitzer Prize winner. Hofstadter introduces ideas from logic, science, and music from oddly appropriate fictional dialogs. Not to everyone's taste but careful reading will pay off.)

Ifrah, G. (1985). *From One to Zero: A Universal History of Numbers*. New York: Viking Penguin. (Another history of numeration, but an excellent one.)

Jacobowitz, H. (1962). *Computer Arithmetic*. New York: Hayden Publishing. (There are many superficial surveys of computer arithmetic. This is one that does a thorough job of presenting the concepts and how they are implemented.)

Livio, M. (2003). *The Story of Phi, the World's Most Astonishing Number*, New York: Broadway Books. (Another irrational number that turns up in remarkable places: from breeding rabbits to tree growth. Too many religious references for my taste but still a good book.)

Maor, E. (1998). *e: The Story of a Number*. Princeton, NJ: Princeton University Press. (The base of the natural logarithms forms the basis of much science: radioactive decay and continuous interest are only two examples. A good story, well told.)

Maor, E. (1998). *Trigonometric Delights*. Princeton, NJ: Princeton University Press. (An accurate title. Well presented examples make this a pleasant read.)

Maxfield, C. and A. Brown (2005). *The Definitive Guide to How Computers Do Math*. Hoboken, NJ: Wiley. (This book takes you to the heart of a computer to show how basic procedures are implemented. You learn how much is going on just to arrive at the four basic operations applied to integer arithmetic.)

Menninger, K. (1969). *Number Words and Number Symbols: A Cultural History of Numbers*. Cambridge, MA: M.I.T. Press. (Another respected wide-ranging survey of mathematical ideas.)

Nahin, P. J. (1998). *An Imaginary Tale: The Story of i*. Princeton, NJ: Princeton University Press. (Another recent book about the "special" numbers of mathematics. Extends the ideas of *Inside Your Calculator*, Chapter 9.)

Nahin, P. J. (2004). *When Least Is Best: How Mathematicians Discovered Many Clever Ways to Make Things as Small (or as Large) as Possible*. Princeton, NJ: Princeton University Press. (Most of mathematics instruction is focused on equations. This book takes another equally important route: to the inequalities. A series of remarkable problems are addressed. Challenging.)

Parris, R. (2001). "Elementary Functions and Calculators." available from `math.exeter.edu/rparris/peanut/cordic.pdf`. (The best careful introduction to CORDIC I have been able to find. Still heavy going, but well worth your attention if you wish to explore this concept further.)

Posamentier, A. S. and I. Lehmann (2004). *π: A Biography of the World's Most Mysterious Number*. Amherst, NY: Prometheus Books. (A straightforward and up-to-date survey of ideas related to π accessible to general readers. If you want to use—or memorize—thousands of digits in the decimal expansion of π, you'll find them here.)

Ralston, A., E. D. Reilly, and D. Hemmendinger, Eds. (2000). *Encyclopedia of Computer Science*. London: Nature Publishing Group. (It is always a compliment to say that a book is encyclopedic in scope. This one certainly earns its name. Much history is included in this excellent resource for all aspects of computing.)

Reid, C. (1964). *From Zero to Infinity: What Makes Numbers Interesting*. New York: Thomas Y. Crowell Company. (Another accessible survey by a fine science historian.)

Stein, S. K. (1999). *Archimedes: What Did He Do Besides Cry Eureka?* Washington, D. C.: Mathematical Association of America. (Sherman Stein is another master expositor. Here he explains the key ideas presented by one of the world's finest mathematicians. A source for much of *Inside Your Calculator* Chapter 7 and Appendixes Q and R.)

Stein, S. K. (1999). *Mathematics: The Man-Made Universe*, New York: Dover Publications. (A new edition of a fine collection of mathematical topics well explained. Any trade book or textbook by Stein is worth studying. (A famous comment about Stein's calculus text: Use this book for teaching, assign to your students anyone else's.))

Stover, D. W. (2006). *Precalculus Problems and Projects*, in preparation. (When published, this book should be on every school and college mathematics teacher's desk. Most of the ideas of *Inside Your Calculator* were derived from Stover's much more extended development of precalculus mathematics.)

Weisstein, E. (1999). "Fermat's Last Theorem," available from `mathworld .wolfram.com/FermatsLastTheorem.html`. (If Fermat really could prove the theorem that bears his name as he claims, he certainly did not take this modern route. Worth exploring if only to see how esoteric is the math involved.)

INDEX

Inside Your Calculator: From Simple Programs to Significant Insights   By Gerald R. Rising
Copyright © 2007 John Wiley & Sons, Inc.

DATE DUE

FEB 17 2012		